The Open University

C5

Fractals

This publication forms part of an Open University course. Details of this and other Open University courses can be obtained from the Student Registration and Enquiry Service, The Open University, PO Box 197, Milton Keynes, MK7 6BJ, United Kingdom: tel. +44 (0)870 333 4340, e-mail general-enquiries@open.ac.uk

Alternatively, you may visit the Open University website at http://www.open.ac.uk where you can learn more about the wide range of courses and packs offered at all levels by The Open University.

To purchase a selection of Open University course materials, visit the webshop at www.ouw.co.uk, or contact Open University Worldwide, Michael Young Building, Walton Hall, Milton Keynes, MK7 6AA, United Kingdom, for a brochure: tel. +44 (0)1908 858785, fax +44 (0)1908 858787, e-mail ouwenq@open.ac.uk

The Open University, Walton Hall, Milton Keynes, MK7 6AA.

First published 2006.

Copyright © 2006 The Open University

All rights reserved; no part of this publication may be reproduced, stored in a retrieval system, transmitted or utilised in any form or by any means, electronic, mechanical, photocopying, recording or otherwise, without written permission from the publisher or a licence from the Copyright Licensing Agency Ltd. Details of such licences (for reprographic reproduction) may be obtained from the Copyright Licensing Agency Ltd, 90 Tottenham Court Road, London W1T 4LP.

Open University course materials may also be made available in electronic formats for use by students of the University. All rights, including copyright and related rights and database rights, in electronic course materials and their contents are owned by or licensed to The Open University, or otherwise used by The Open University as permitted by applicable law.

In using electronic course materials and their contents you agree that your use will be solely for the purposes of following an Open University course of study or otherwise as licensed by The Open University or its assigns.

Except as permitted above you undertake not to copy, store in any medium (including electronic storage or use in a website), distribute, transmit or re-transmit, broadcast, modify or show in public such electronic materials in whole or in part without the prior written consent of The Open University or in accordance with the Copyright, Designs and Patents Act 1988.

Edited, designed and typeset by The Open University, using the Open University TeX System.

Printed and bound in the United Kingdom by The Charlesworth Group, Wakefield.

ISBN 0 7492 4139 X

Contents

Introduction	**4**
Study guide	4
1 Examples of fractals	**5**
1.1 Cantor sets	5
1.2 The Sierpiński gasket	9
1.3 The von Koch curve	11
2 Using the Contraction Mapping Theorem	**16**
2.1 The Hausdorff metric	16
2.2 Constructing contraction mappings	22
2.3 Iterated function schemes	25
2.4 Proof of Theorem 2.5	30
3 Dimensions of fractals	**33**
3.1 Box dimension	33
3.2 The open set condition	37
Conclusion	**46**
Solutions to problems	**47**
Index	**52**

Introduction

In this, the final unit of the course, we show how the theory of topological spaces is used to great effect in the study of fractals, an area where there is much current interest and research activity. You have probably seen computer-generated pictures of fractals and may even have produced such pictures yourself. One picture you may have seen is the *von Koch snowflake* (Figure 0.1).

There is no standard definition of a fractal, but there are certain characteristics that we expect a fractal to have. The von Koch snowflake is typical of many fractals in the following ways:

▶ it has a fine structure — the closer you look at it, the more detail you see;

▶ it has some degree of self-similarity — the same patterns keep appearing;

▶ it is difficult to describe in terms of classical geometry.

von Koch snowflake

Figure 0.1

Unit C4, Section 3.

Despite having an intricate structure, many fractals can be defined very simply. In this unit we use the Contraction Mapping Theorem of *Unit C4* to generate a large class of fractals, including the von Koch snowflake.

We also look at the *dimensions* of fractals. Many fractals cannot be described adequately in terms of the usual integer-valued definitions of dimension. For example, the von Koch snowflake does not have finite length (suggesting that it has dimension greater than 1) and has zero area (suggesting that it has dimension less than 2). We introduce a definition of dimension that gives a good indication of the 'size' of many fractals, but which can take non-integer values.

Study guide

In Section 1, *Examples of fractals*, we look at some well-known examples of fractals and investigate their properties. You need to become familiar with these examples since they appear throughout the unit.

In Section 2, *Using the Contraction Mapping Theorem*, you will see how the Contraction Mapping Theorem gives a method for generating a large class of fractals. This is the longest section of the unit, and you may wish to spend more than one study session on it. Section 2.4 is not assessed.

In Section 3, *Dimensions of fractals*, we show a way of defining the dimension of a fractal.

There is software associated with this unit. It enables you to generate pictures of fractals by using the ideas introduced in Section 2, and you will be able to use it at any time after studying this section.

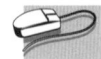

1 Examples of fractals

> After working through this section, you should be able to:
> ▶ describe the *middle-third Cantor set*, the *Sierpiński gasket* and the *von Koch snowflake*;
> ▶ explain what is meant by a *similarity*;
> ▶ determine the *ratio* of a similarity;
> ▶ identify *self-similar* sets and the similarities defining them.

In this section we look at some well-known examples of fractals. By studying their properties, you should get a feeling for the characteristics that are typical of fractals in general.

1.1 Cantor sets

We begin by looking at a class of fractals known as *Cantor sets*. These are subsets of the real line that are obtained by systematically removing open intervals from a closed interval, to obtain a compact, totally disconnected set. The best-known of these is the *middle-third Cantor set* that you met earlier (see Figure 1.1).

Unit C2, Example 3.2.

Figure 1.1 Middle-third Cantor set

You may recall that we obtain the middle-third Cantor set by a simple recursive procedure. We begin with the line segment $K_0 = [0, 1]$.

Figure 1.2

We then remove the open middle-third of K_0 — the open interval $(\frac{1}{3}, \frac{2}{3})$ — leaving a set K_1 made up of two closed intervals, each of length one-third. So

$$K_1 = \left[0, \tfrac{1}{3}\right] \cup \left[\tfrac{2}{3}, 1\right].$$

This explains the name *middle-third* Cantor set.

Figure 1.3

We then apply the same procedure to each of the two line segments in K_1 to give the set

$$K_2 = \left[0, \tfrac{1}{9}\right] \cup \left[\tfrac{2}{9}, \tfrac{1}{3}\right] \cup \left[\tfrac{2}{3}, \tfrac{7}{9}\right] \cup \left[\tfrac{8}{9}, 1\right].$$

Figure 1.4

We continue in this way, constructing a set K_n for each $n \in \mathbb{N}$. Thus K_n is formed by removing the open middle-third of each line segment in K_{n-1}. The set K_n consists of 2^n closed intervals, each with length $\frac{1}{3^n}$.

The middle-third Cantor set is the intersection of the sets K_n. We denote this set by

$$K_C = \bigcap_{n \in \mathbb{N}} K_n.$$

You will see that many fractals can be defined by a similar recursive method.

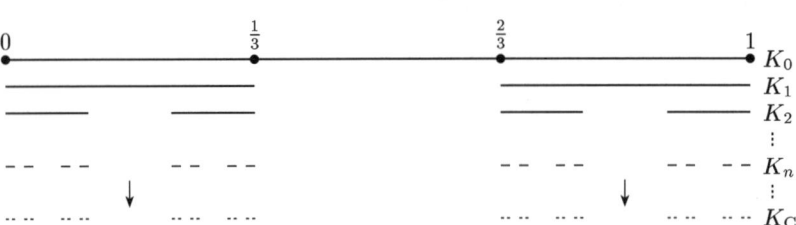

Figure 1.5

Each set K_n is closed and bounded, and hence compact. Since $K_{n+1} \subseteq K_n$ for each $n \in \mathbb{N}$, the Cantor set K_C is the intersection of a nested sequence of non-empty compact sets and is therefore non-empty and compact. In fact, K_C contains the endpoints of each of the 2^n line segments forming K_n, for each $n \in \mathbb{N}$, since none of these endpoints is removed in the recursive procedure. Thus K_C is an infinite set.

Unit C2, Theorem 3.10.

In fact, K_C contains uncountably many points.

Problem 1.1

For each $n \in \mathbb{N}$, calculate the total length L_n of the line segments that form K_n, and find $\lim_{n \to \infty} L_n$.

Since K_C is a subset of K_n for each $n \in \mathbb{N}$, the solution to Problem 1.1 suggests that K_C has 'zero length', and is therefore 'smaller' than a 1-dimensional set (such as a line segment). It is also 'larger' than a 0-dimensional set (such as a point), since it contains infinitely many points. In Section 3, we give a definition of dimension that allows us to assign K_C a dimension between 0 and 1.

Self-similarity

In the construction of the middle-third Cantor set, the set K_1 is made up of two copies of K_0, each scaled by $\frac{1}{3}$.

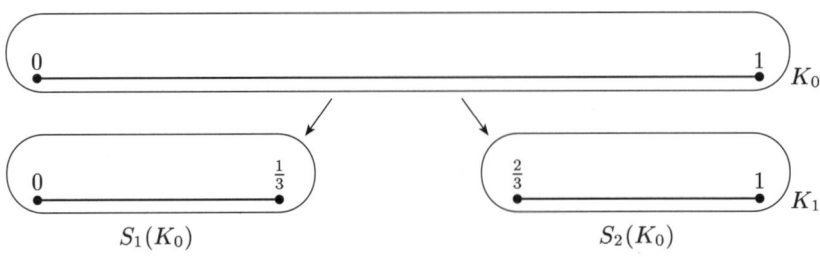

Figure 1.6

Similarly, the set K_2 consists of two copies of K_1, each scaled by $\frac{1}{3}$ — and in general, for each $n \in \mathbb{N}$, the set K_{n+1} consists of two copies of K_n, each scaled by $\frac{1}{3}$.

The mappings taking K_n to these two smaller copies of itself are the functions $S_1, S_2 \colon \mathbb{R} \to \mathbb{R}$, defined by

$$S_1(x) = \tfrac{1}{3}x \quad \text{and} \quad S_2(x) = \tfrac{1}{3}(x+2).$$

The mapping S_1 scales the Cantor set by $\tfrac{1}{3}$, and the mapping S_2 scales the Cantor set by $\tfrac{1}{3}$ and translates the resulting set $\tfrac{2}{3}$ to the right.

Each of the mappings S_1 and S_2 is a *similarity*.

> **Definition**
>
> A mapping $S \colon \mathbb{R}^m \to \mathbb{R}^m$ is a **similarity** if there is a $c > 0$ such that
>
> $$d^{(m)}(S(\mathbf{x}), S(\mathbf{y})) = c\,d^{(m)}(\mathbf{x}, \mathbf{y}) \quad \text{for all } \mathbf{x}, \mathbf{y} \in \mathbb{R}^m.$$
>
> The constant c is the **similarity ratio**.

Each of the functions S_1 and S_2 scales by $\tfrac{1}{3}$, so has similarity ratio $\tfrac{1}{3}$. Any translation, reflection or rotation has similarity ratio 1, because all the distances remain unchanged.

A similarity from $\mathbb{R}^2$ to $\mathbb{R}^2$ has the form

$$S(\mathbf{x}) = c\mathbf{A}\mathbf{x}^\mathrm{T} + \mathbf{y},$$

where $\mathbf{A}$ is a 2×2 orthogonal matrix, $\mathbf{y}$ is a translation vector, and c is the similarity ratio. In particular, if S involves an anticlockwise rotation by θ about the origin, then

$$\mathbf{A} = \begin{pmatrix} \cos\theta & -\sin\theta \\ \sin\theta & \cos\theta \end{pmatrix}.$$

Here $\mathbf{x}^\mathrm{T}$ denotes the transpose of $\mathbf{x}$.

$\mathbf{A}$ is *orthogonal* if $\mathbf{A}\mathbf{A}^\mathrm{T} = \mathbf{I}$.

Problem 1.2

Write down the similarity ratio of each of the following mappings.

(a) $S \colon \mathbb{R} \to \mathbb{R}$ defined by $S(x) = \tfrac{1}{2}x + 3$.

(b) $S \colon \mathbb{R}^2 \to \mathbb{R}^2$ defined by $S((x,y)) = (\tfrac{1}{4}y, \tfrac{1}{4}(x+3))$.

For the similarities S_1 and S_2 used to define the Cantor set, each set K_n in the construction of K_C satisfies

$$K_{n+1} = S_1(K_n) \cup S_2(K_n).$$

Since K_C is a 'limit' of the sets K_n, we may expect that

$$K_\mathrm{C} = S_1(K_\mathrm{C}) \cup S_2(K_\mathrm{C}).$$

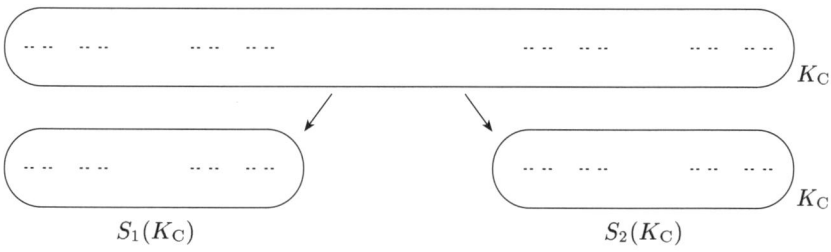

Figure 1.7

In Section 2, we show that this is indeed the case. We say that the middle-third Cantor set is *self-similar*.

> **Definition**
>
> A set $A \subseteq \mathbb{R}^m$ is **self-similar** if there are similarities $S_i \colon \mathbb{R}^m \to \mathbb{R}^m$ with similarity ratios $0 < c_i < 1$, for $1 \leq i \leq k$, such that
> $$A = \bigcup_{i=1}^{k} S_i(A).$$

The set A consists of k copies of itself, each of which is the image of A under a similarity.

The self-similarity of the middle-third Cantor set means that it is regular, with the same patterns (but smaller) appearing again and again the closer you look at the set.

Most fractals have some degree of self-similarity. There are also many simple sets that are self-similar.

Problem 1.3

Show that the line segment $[0, 1]$ is self-similar, by writing down two similarities $S_1, S_2 \colon \mathbb{R} \to \mathbb{R}$, with similarity ratios $0 < c_1 < 1$ and $0 < c_2 < 1$, such that
$$[0, 1] = S_1([0, 1]) \cup S_2([0, 1]).$$

Not all simple sets are self-similar; for example, it can be shown that no circle is self-similar.

Many other fractals can be obtained by a procedure similar to that used to obtain the middle-third Cantor set. We now ask you to investigate one of these.

Problem 1.4

Let $K_0 = [0, 1]$. Remove the second and fourth quarters of K_0 to give a new set
$$K_1 = [0, \tfrac{1}{4}] \cup [\tfrac{1}{2}, \tfrac{3}{4}].$$

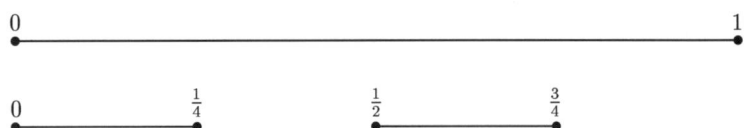

Figure 1.8

For each $n \in \mathbb{N}$, apply this procedure to each closed line segment in K_n to give a new set K_{n+1}.

(a) Draw the set K_2.

(b) For each $n \in \mathbb{N}$, calculate the total length L_n of the line segments that form K_n, and find $\lim_{n \to \infty} L_n$.

(c) Write down similarities $S_1, S_2 \colon \mathbb{R} \to \mathbb{R}$ such that
$$K_{n+1} = S_1(K_n) \cup S_2(K_n) \quad \text{for each } n \in \mathbb{N},$$
and state their similarity ratios.

1.2 The Sierpiński gasket

Our next example of a fractal is the *Sierpiński gasket* (Figure 1.9).

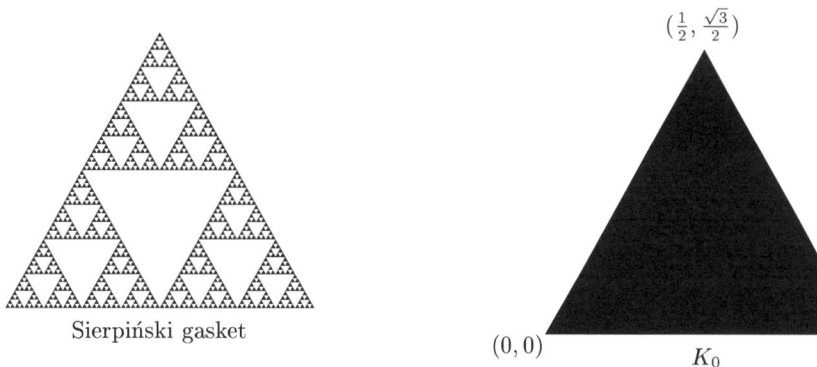

Figure 1.9

Figure 1.10

This fractal is obtained by a recursive procedure. We begin (Figure 1.10) with an upright closed equilateral triangle K_0 whose base is the line segment joining the points $(0,0)$ and $(1,0)$.

We then remove an open equilateral triangle from K_0, leaving a set K_1 made up of three closed equilateral triangles, each with height half that of K_0 (Figure 1.11).

Wacław Sierpiński (1882–1969) was one of the founders of the Polish school of mathematics, which focused on point-set topology, real analysis and logic. He published more than 600 papers, and introduced a number of thought-provoking sets, including the 'gasket' that is named after him.

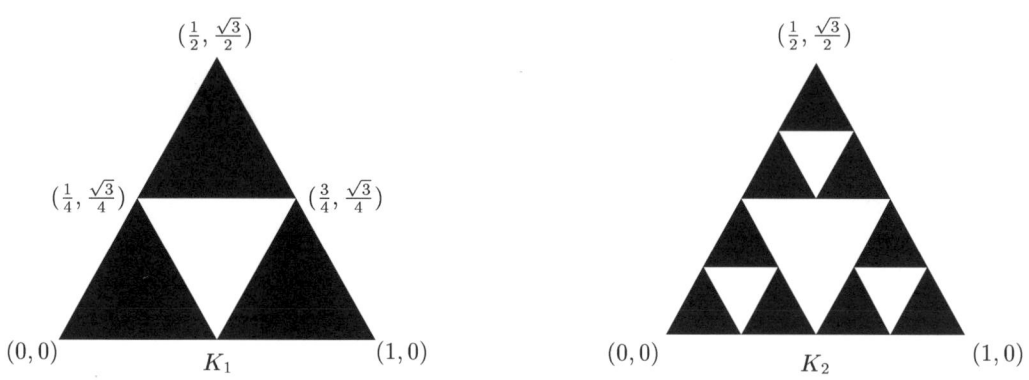

Figure 1.11

Figure 1.12

We apply this procedure to each of the three triangles in K_1 to give a new set K_2 (Figure 1.12).

We continue in this way, constructing a set K_n for each $n \in \mathbb{N}$. The set K_n is formed from 3^n equilateral triangles, each with height $1/2^n$ times the height of K_0. The Sierpiński gasket K_G is defined to be the intersection of the sets K_n,

$$K_G = \bigcap_{n \in \mathbb{N}} K_n.$$

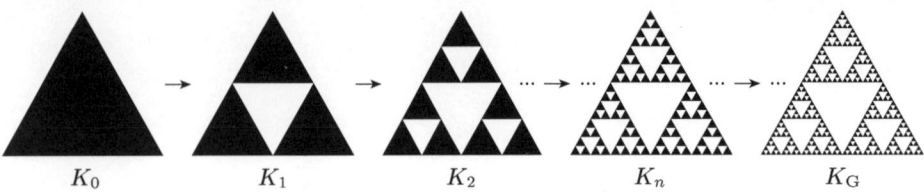

Figure 1.13

Each set K_n is closed and bounded, and hence compact. Since $K_{n+1} \subseteq K_n$, for each $n \in \mathbb{N}$, the Sierpiński gasket K_G is the intersection of a nested sequence of non-empty compact sets, and is therefore non-empty and compact. In fact, K_G contains the boundary of each of the 3^n equilateral triangles forming K_n, for each $n \in \mathbb{N}$, since no boundary points are removed in the recursive procedure (Figure 1.13).

Unit C2, Theorem 3.10.

Problem 1.5

For each $n \in \mathbb{N}$, calculate the total length of the boundaries of the 3^n equilateral triangles that form K_n. Deduce that the Sierpiński gasket K_G does not have finite length.

We now ask you to show that K_G has 'zero area'. This is more surprising, since each set K_n has positive area.

Problem 1.6

For each $n \in \mathbb{N}$, calculate the area of K_n. Deduce that the Sierpiński gasket has zero area.

The Sierpiński gasket seems to be smaller than a 2-dimensional set (such as a rectangle), since it has zero area, and yet it seems to be larger than a 1-dimensional set (such as a line segment), since it is bounded but does not have finite length.

We address this issue in Section 3.

Note that the Sierpiński gasket appears to be self-similar — it appears to be made up of three copies of itself, each with height $\tfrac{1}{2}$ times the height of the whole gasket.

Problem 1.7

Let K_0 and K_1 be the first two sets in the construction of the Sierpiński gasket. Write down three similarities $S_1, S_2, S_3 \colon \mathbb{R}^2 \to \mathbb{R}^2$ such that

$$K_1 = S_1(K_0) \cup S_2(K_0) \cup S_3(K_0),$$

and state their similarity ratios.

For three such similarities S_1, S_2 and S_3,

$$K_{n+1} = S_1(K_n) \cup S_2(K_n) \cup S_3(K_n),$$

for each $n \in \mathbb{N}$, and this suggests that

$$K_G = S_1(K_G) \cup S_2(K_G) \cup S_3(K_G).$$

We justify this in Section 2.

Many other fractals can be obtained by a procedure similar to that used to obtain the Sierpiński gasket.

Problem 1.8

Let K_0 be the closed square with corners at the points $(0,0)$, $(0,1)$, $(1,0)$ and $(1,1)$. Divide K_0 into nine subsquares and remove all but the middle and corner closed subsquares to give a new set K_1.

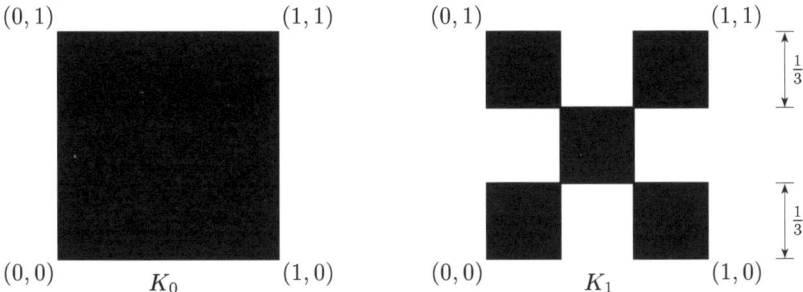

Figure 1.14

For each $n \in \mathbb{N}$, apply this procedure to each of the closed squares in K_n to give a new set K_{n+1}. Let K denote the intersection of the sets K_n.

(a) Draw the set K_2.

(b) Show that K is non-empty and compact and has zero area.

(c) Write down five similarities $S_1, S_2, \ldots, S_5 \colon \mathbb{R}^2 \to \mathbb{R}^2$ such that

$$K_{n+1} = \bigcup_{i=1}^{5} S_i(K_n) \quad \text{for each } n \in \mathbb{N},$$

and state their similarity ratios.

1.3 The von Koch curve

We conclude this section by returning to the von Koch snowflake that you met in the Introduction. The snowflake is made up of three copies of the von Koch curve, as shown in Figure 1.15.

The Swedish mathematician Helge von Koch (1870–1924) described this curve in 1906, showing that no point on the curve has a tangent.

von Koch curve

Figure 1.15

Although the von Koch curve appears to be very intricate, it is obtained by a simple procedure. We begin with the line segment $K_0 = [0, 1] \times \{0\}$ (Figure 1.16).

Figure 1.16

We then remove the open middle-third of K_0 and replace it with the other two sides of the equilateral triangle based on the missing segment. This gives a set K_1 (Figure 1.17).

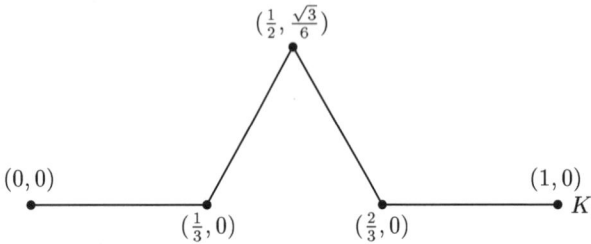

Figure 1.17

We then apply this procedure to each of the four line segments of K_1 to give a new set K_2 (Figure 1.18).

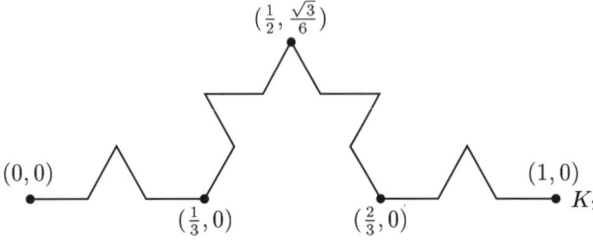

Figure 1.18

We continue in this way, constructing sets K_n for each $n \in \mathbb{N}$. We wish to define the von Koch curve K_V to be the limit of the sets K_n as $n \to \infty$.

At this point it is not clear how to define the *limit* of the sets K_n as $n \to \infty$. In each of our previous examples, we had a nested sequence of compact sets, and so could define the limit set to be the intersection of these nested sets — this must be a non-empty compact set. In this case, the sets K_n are compact, but they are not nested, and it is not clear that a well-defined limit exists. In the next section we show that this limit does exist.

Although it is not obvious how to define the limit of the sets K_n, we can identify certain points that must belong to the von Koch curve. For each $n \in \mathbb{N}$, the endpoints of each line segment in K_n are not removed by the subsequent stages of the construction, and so each of these endpoints belongs to K_V.

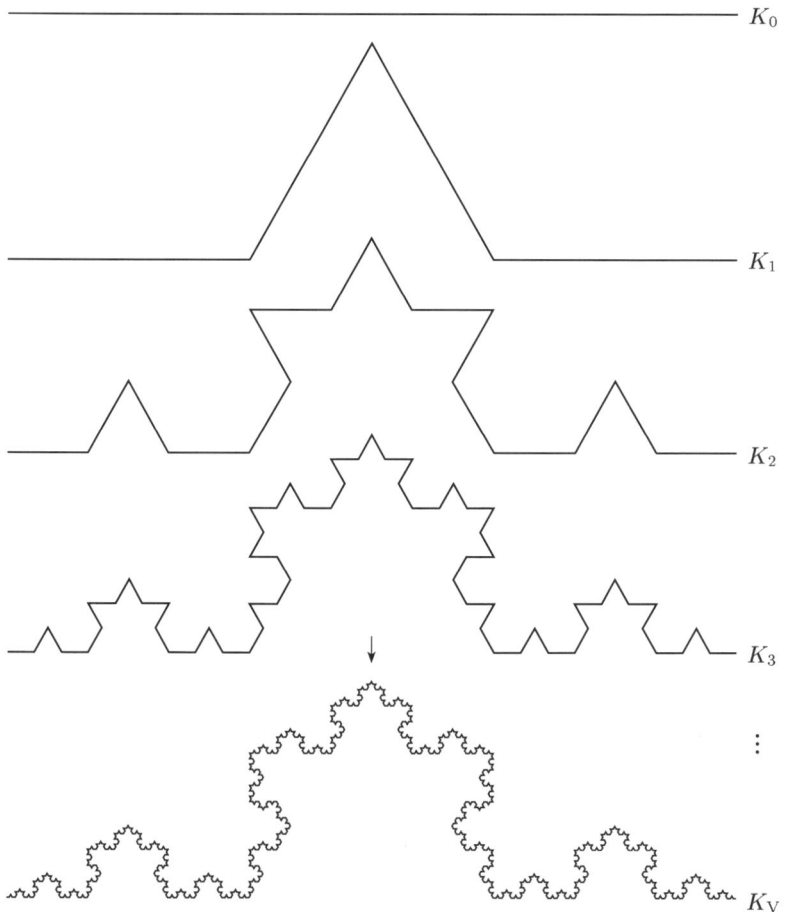

Figure 1.19

Problem 1.9

For each $n \in \mathbb{N}$, determine the length of each line segment that makes up K_n.

Using the fact that the length of the section of the von Koch curve lying between the two endpoints of one of these line segments cannot be less than the distance between the two endpoints, deduce that K_V does not have finite length.

We have just seen that the von Koch curve is very different from most familiar curves. It lies within a bounded part of the plane, yet it is so 'wiggly' that it does not have finite length. The 'wiggly' nature of the curve also means that no point on the curve has a tangent. It is thus difficult to describe the curve in terms of classical geometric shapes.

This is typical of fractals.

For each $n \in \mathbb{N}$, the set K_{n+1} is made up of four copies of K_n, each scaled by $\frac{1}{3}$. So

$$K_{n+1} = S_1(K_n) \cup S_2(K_n) \cup S_3(K_n) \cup S_4(K_n),$$

where the mappings S_1, S_2, S_3, S_4 can be described as in Figures 1.20–1.23.

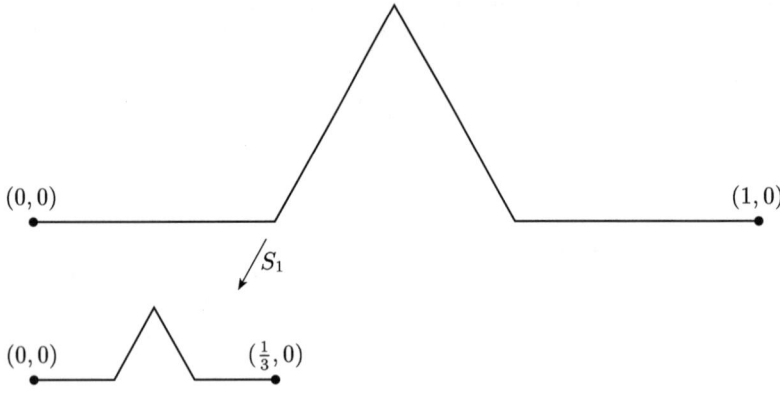

Figure 1.20 S_1 *scales by a factor of* $\frac{1}{3}$.

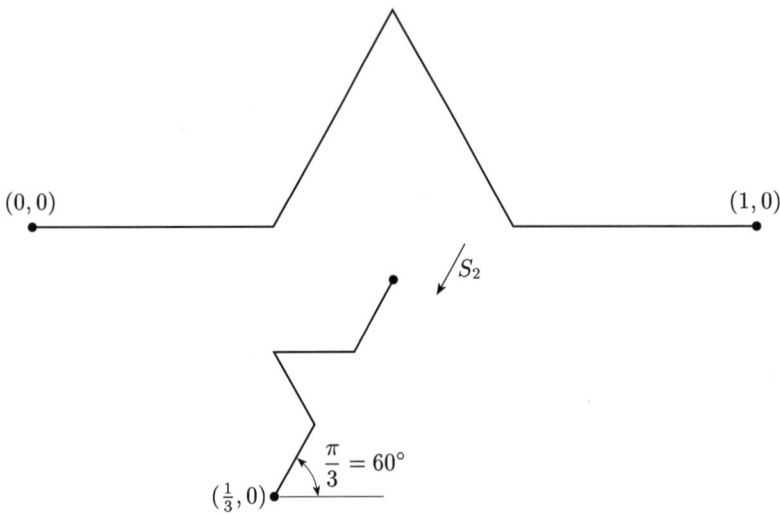

Figure 1.21 S_2 *scales by a factor of* $\frac{1}{3}$, *rotates anticlockwise by* $\frac{1}{3}\pi$ *about the point* $(0,0)$, *and translates by* $(\frac{1}{3}, 0)$.

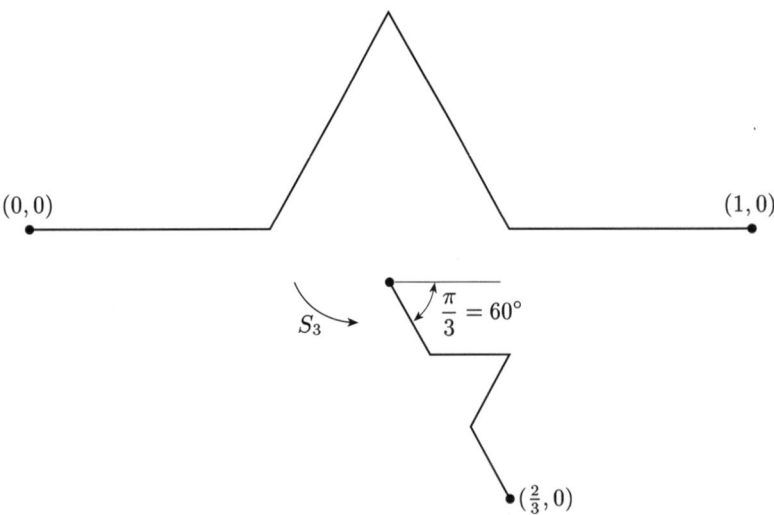

Figure 1.22 S_3 *scales by a factor of* $\frac{1}{3}$, *rotates clockwise by* $\frac{1}{3}\pi$ *about the point* $(0,0)$, *and translates by* $(\frac{1}{2}, \frac{\sqrt{3}}{6})$.

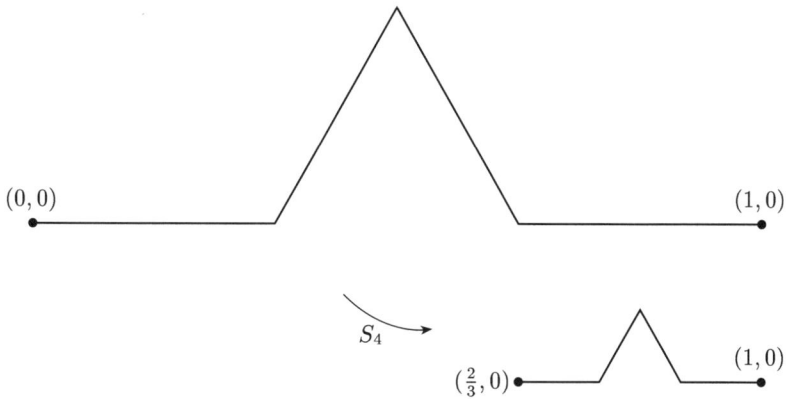

Figure 1.23 S_4 *scales by a factor of* $\frac{1}{3}$ *and translates by* $(\frac{2}{3}, 0)$.

This suggests that if K_V is the limit of the sets K_n, then

$$K_V = S_1(K_V) \cup S_2(K_V) \cup S_3(K_V) \cup S_4(K_V).$$

This is justified in Section 2.

So the von Koch curve appears to be self-similar.

You have now met some well-known fractals — the middle-third Cantor set, the Sierpiński gasket, and the von Koch curve. We will refer to these examples throughout the unit.

2 Using the Contraction Mapping Theorem

After working through this section, you should be able to:
▶ determine the Hausdorff distance between two compact sets;
▶ explain why, for each finite family of contraction mappings, there is a unique non-empty compact set that is invariant for that family;
▶ construct simple examples of invariant sets.

In the previous section you met several examples of fractals, each of which appears to be self-similar. In other words, for each such fractal K, there appears to be a finite family of similarities, $S_1, S_2, \ldots, S_k$, such that

$$K = S_1(K) \cup S_2(K) \cup \cdots \cup S_k(K) = \bigcup_{i=1}^{k} S_i(K).$$

Each fractal K is constructed by taking a compact starting set K_0 and repeatedly applying the similarities $S_1, S_2, \ldots, S_k$.

In this section we use the Contraction Mapping Theorem to show that if we take *any* compact set K_0 and *any* family of contraction mappings $S_1, S_2, \ldots, S_k$, then this recursive procedure defines a set K satisfying $K = \bigcup_{i=1}^{k} S_i(K)$. By looking at different families of contraction mappings, we can construct many different fractals. Note that all the similarities in Section 1 are in fact contraction mappings.

2.1 The Hausdorff metric

We begin by reminding you of the Contraction Mapping Theorem.

Unit C4, Theorem 3.2.

The Contraction Mapping Theorem

Let (X, d) be a complete metric space, and let $T: X \to X$ be a contraction mapping. Then there is a unique element $x_T \in X$ such that

$$T(x_T) = x_T.$$

Moreover, for each $x \in X$, the sequence of iterates

$$x, T(x), T^2(x), T^3(x), \ldots$$

converges to this unique fixed point x_T — that is, for any $x \in X$,

$$d(T^n(x), x_T) \to 0 \text{ as } n \to \infty.$$

We shall use this result to generate a large class of fractals, each of which is the fixed point x_T of a suitable contraction mapping T. We begin by defining a suitable metric space (X, d).

In each of our earlier examples, we repeatedly applied mappings to a *compact* subset of $\mathbb{R}^m$ (where m was 1 or 2). Since this process generated fractals that are also compact subsets of $\mathbb{R}^m$, it seems reasonable to take our set X to be a set whose members are compact subsets of $\mathbb{R}^m$.

Recall that the compact subsets of $\mathbb{R}^m$ are those that are closed and bounded.

> **Definition**
>
> For each $m \in \mathbb{N}$, the set $\mathcal{K}(\mathbb{R}^m)$ is defined as follows:
>
> $\mathcal{K}(\mathbb{R}^m) = \{K \subseteq \mathbb{R}^m : K \text{ is non-empty and compact}\}.$

Each member of $\mathcal{K}(\mathbb{R}^m)$ is itself a set.

We are usually interested in the sets $\mathcal{K}(\mathbb{R})$ and $\mathcal{K}(\mathbb{R}^2)$. For example, the middle-third Cantor set is a set in $\mathcal{K}(\mathbb{R})$ that is generated by repeatedly applying mappings to sets in $\mathcal{K}(\mathbb{R})$, and the Sierpiński gasket is a set in $\mathcal{K}(\mathbb{R}^2)$ that is generated by repeatedly applying mappings to sets in $\mathcal{K}(\mathbb{R}^2)$.

Problem 2.1

Which of the following sets belong to $\mathcal{K}(\mathbb{R})$?

(a) $(-1, 1]$ (b) $[0, 2]$ (c) $[0, \infty)$ (d) $\{1\}$ (e) $[0, 2] \cap [4, 6]$

In order to apply the Contraction Mapping Theorem, we need to define a *metric* on the set $\mathcal{K}(\mathbb{R}^m)$ — that is, a distance between any two sets in $\mathcal{K}(\mathbb{R}^m)$ for which the metric properties (M1)–(M3) are satisfied. We begin by considering the simpler problem of how to define the distance between a *point* in $\mathbb{R}^m$ and a set K in $\mathcal{K}(\mathbb{R}^m)$. It seems reasonable to define this as follows.

> **Definition**
>
> Let $\mathbf{x} \in \mathbb{R}^m$ and let $K \in \mathcal{K}(\mathbb{R}^m)$, for some $m \in \mathbb{N}$.
>
> The **Euclidean distance** between the point $\mathbf{x}$ and the set K is
>
> $d^{(m)}(\mathbf{x}, K) = \min\{d^{(m)}(\mathbf{x}, \mathbf{k}) : \mathbf{k} \in K\}.$

Remark

This definition makes sense since the function $d^{(m)}(\mathbf{x}, \mathbf{k})$ is a *continuous* function of $\mathbf{k}$ on the *compact* set K. It follows from the General Extreme Value Theorem that there is a $\mathbf{c} \in K$ such that

Unit A1, Worked problem 5.2.
Unit C2, Theorem 4.3.

$d^{(m)}(\mathbf{x}, \mathbf{c}) \leq d^{(m)}(\mathbf{x}, \mathbf{k}) \quad \text{for each } \mathbf{k} \in K.$

So

$d^{(m)}(\mathbf{x}, K) = d^{(m)}(\mathbf{x}, \mathbf{c}).$

For example, consider the point $\mathbf{x} = (3, 0)$ and the set $K = \{(x, y) : |x| \leq 2, |y| \leq 1\}$, which is a member of $\mathcal{K}(\mathbb{R}^2)$.

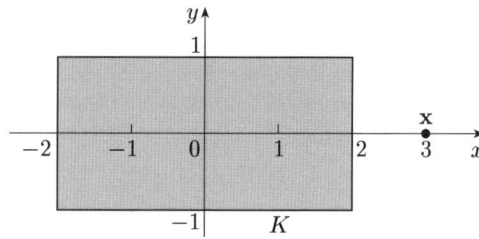

Figure 2.1

The closest point in K to the point $(3, 0)$ is the point $(2, 0)$, and so

$d^{(2)}((3, 0), K) = d^{(2)}((3, 0), (2, 0)) = 1.$

Problem 2.2

Determine the following.

(a) $d^{(1)}(1, [2, 3])$ (b) $d^{(1)}(1, [0, 2])$

(c) $d^{(2)}((0, 3), \{(x, y) : x^2 + y^2 \leq 1\})$ (d) $d^{(2)}((2, 1), \{(3, y) : |y| \leq 1\})$

Note that $d^{(m)}(\mathbf{x}, K) = 0$ precisely when there is a point $\mathbf{c} \in K$ such that $d^{(m)}(\mathbf{x}, \mathbf{c}) = 0$ — that is, $\mathbf{x} = \mathbf{c}$. This gives the following result.

> **Lemma 2.1**
>
> Let $\mathbf{x} \in \mathbb{R}^m$ and let $K \in \mathcal{K}(\mathbb{R}^m)$, for some $m \in \mathbb{N}$. Then
>
> $d^{(m)}(\mathbf{x}, K) = 0$ if and only if $\mathbf{x} \in K$.

The following lemma gives a useful property of $d^{(m)}$ for compact sets.

> **Lemma 2.2**
>
> Let $\mathbf{x} \in \mathbb{R}^m$ and let $J, K \in \mathcal{K}(\mathbb{R}^m)$, for some $m \in \mathbb{N}$. Let $J \subseteq K$; then
>
> $d^{(m)}(\mathbf{x}, K) \leq d^{(m)}(\mathbf{x}, J)$.

Problem 2.3

Prove Lemma 2.2.

Having established this definition of the distance from a point to a compact set, we now extend it to the definition of the distance from one compact set to another as follows.

> **Definition**
>
> Let $J, K \in \mathcal{K}(\mathbb{R}^m)$, for some $m \in \mathbb{N}$. The **Euclidean distance** between the set J and the set K is
>
> $d^{(m)}(J, K) = \max\{d^{(m)}(\mathbf{j}, K) : \mathbf{j} \in J\}$.

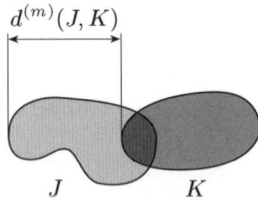

Figure 2.2

Remark

We use 'max' here instead of 'min' as 'min' would give a distance of zero between the pair of sets shown in Figure 2.3.

We want distinct sets to be a positive distance apart, otherwise we do not have a metric.

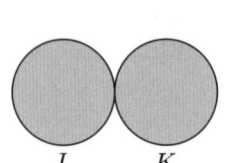

Figure 2.3

This definition makes sense since the function $d^{(m)}(\mathbf{j}, K)$ is a *continuous* function of $\mathbf{j}$ on the *compact* set J. It follows from the General Extreme Value Theorem (Theorem 4.3 of *Unit C2*) that there is a $\mathbf{d} \in J$ such that

$d^{(m)}(\mathbf{j}, K) \leq d^{(m)}(\mathbf{d}, K)$ for each $\mathbf{j} \in J$.

In the Exercises, we ask you to prove that $d^{(m)}(\mathbf{j}, K)$ is a continuous function of $\mathbf{j}$.

So

$d^{(m)}(J, K) = d^{(m)}(\mathbf{d}, K)$.

Since there is a $\mathbf{c} \in K$ such that $d^{(m)}(\mathbf{d}, K) = d^{(m)}(\mathbf{d}, \mathbf{c})$, it follows that

$d^{(m)}(J, K) = d^{(m)}(\mathbf{d}, \mathbf{c})$.

In other words, the Euclidean distance between two non-empty compact sets J and K is realized by a pair of points from J and K.

Worked problem 2.1

Let
$$J = \{(3, y) : |y| \leq 1\} \quad \text{and} \quad K = \{(x, y) : |x| \leq 2, |y| \leq 1\}.$$
Determine $d^{(2)}(J, K)$ and $d^{(2)}(K, J)$.

Solution

We begin by sketching the sets J and K.

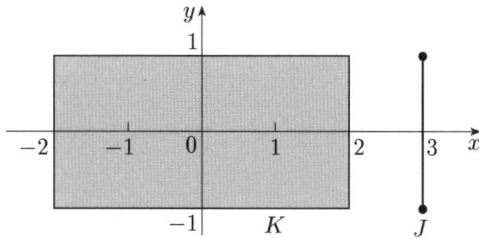

Figure 2.4

The points in J are all at distance 1 from the set K; for example,
$$d^{(2)}(J, K) = d^{(2)}((3, 0), K) = d^{(2)}((3, 0), (2, 0)) = 1.$$

The points in K that are furthest from the set J are those on the left-hand edge of K. One such point is $(-2, 0)$, and so
$$d^{(2)}(K, J) = d^{(2)}((-2, 0), J) = d^{(2)}((-2, 0), (3, 0)) = 5. \qquad \blacksquare$$

Problem 2.4

Let
$$J = \{(x, y) : 0 \leq x^2 + y^2 \leq 1\} \quad \text{and} \quad K = \{(1, 0)\}.$$
Determine $d^{(2)}(J, K)$ and $d^{(2)}(K, J)$.

These examples show that the Euclidean distance from one non-empty compact set to another is *not* a metric, since the symmetry property (M2) is not satisfied. In fact (M1) is not satisfied either, as the first part of the following problem shows.

You may suspect that we have chosen a poor definition of distance between compact sets, but you will see shortly how to rescue this definition.

Problem 2.5

Let $J, K \in \mathcal{K}(\mathbb{R}^m)$, for some $m \in \mathbb{N}$, and let $J \subseteq K$.

(a) Show that $d^{(m)}(J, K) = 0$.

(b) Show that $d^{(m)}(L, J) \geq d^{(m)}(L, K)$, for each $L \in \mathcal{K}(\mathbb{R}^m)$.

We now show that the Euclidean distance *does* satisfy the Triangle Inequality — property (M3). This suggests that, by modifying the definition of this distance slightly, we may be able to obtain a definition of distance between two compact sets that is a metric, as we require.

Lemma 2.3

Let $J, K, L \in \mathcal{K}(\mathbb{R}^m)$, for some $m \in \mathbb{N}$. Then
$$d^{(m)}(J, L) \leq d^{(m)}(J, K) + d^{(m)}(K, L).$$

Proof We know that $d^{(m)}$ is a metric on $\mathbb{R}^m$.

By (M3), for each $\mathbf{j} \in J$, $\mathbf{k} \in K$, $\mathbf{l} \in L$,
$$d^{(m)}(\mathbf{j},\mathbf{l}) \leq d^{(m)}(\mathbf{j},\mathbf{k}) + d^{(m)}(\mathbf{k},\mathbf{l}).$$
Taking $\mathbf{l} \in L$ such that $d^{(m)}(\mathbf{k},\mathbf{l}) = d^{(m)}(\mathbf{k},L)$, we obtain
$$d^{(m)}(\mathbf{j},\mathbf{l}) \leq d^{(m)}(\mathbf{j},\mathbf{k}) + d^{(m)}(\mathbf{k},L).$$
Taking $\mathbf{k} \in K$ such that $d^{(m)}(\mathbf{j},\mathbf{k}) = d^{(m)}(\mathbf{j},K)$, we obtain
$$d^{(m)}(\mathbf{j},\mathbf{l}) \leq d^{(m)}(\mathbf{j},K) + d^{(m)}(\mathbf{k},L).$$
Since $d^{(m)}(\mathbf{j},L) \leq d^{(m)}(\mathbf{j},\mathbf{l})$ and $d^{(m)}(\mathbf{k},L) \leq d^{(m)}(K,L)$, we have
$$d^{(m)}(\mathbf{j},L) \leq d^{(m)}(\mathbf{j},K) + d^{(m)}(K,L).$$
Finally, taking $\mathbf{j} \in J$ such that $d^{(m)}(\mathbf{j},L) = d^{(m)}(J,L)$, we obtain
$$d^{(m)}(J,L) \leq d^{(m)}(\mathbf{j},K) + d^{(m)}(K,L)$$
$$\leq d^{(m)}(J,K) + d^{(m)}(K,L). \qquad \blacksquare$$

So the Euclidean distance from one compact set to another satisfies property (M3) of a metric space, but does not satisfy properties (M1) and (M2). In order to overcome this problem, we introduce the following definition of the distance between two compact sets.

Definition

Let $J, K \in \mathcal{K}(\mathbb{R}^m)$, for some $m \in \mathbb{N}$. The **Hausdorff distance** d_H between the sets J and K is
$$d_\mathrm{H}(J,K) = \max\{d^{(m)}(J,K), d^{(m)}(K,J)\}.$$

For example, consider the sets
$$J = \{(3,y) : |y| \leq 1\} \quad \text{and} \quad K = \{(x,y) : |x| \leq 2, |y| \leq 1\}.$$
In Worked problem 2.1 we showed that $d^{(2)}(J,K) = 1$ and $d^{(2)}(K,J) = 5$. So $d_\mathrm{H}(J,K) = d_\mathrm{H}(K,J) = \max\{1,5\} = 5$.

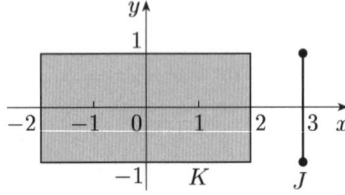

Figure 2.5

Problem 2.6

Let
$$J = \{(x,y) : 0 \leq x^2 + y^2 \leq 1\} \quad \text{and} \quad K = \{(1,0)\}.$$
Determine $d_\mathrm{H}(J,K)$.

The definition of Hausdorff distance clearly satisfies property (M2) of a metric space, and we now show that it is in fact a metric on $\mathcal{K}(\mathbb{R}^m)$.

Theorem 2.4

The Hausdorff distance, d_H, is a metric on $\mathcal{K}(\mathbb{R}^m)$.

Proof Recall that, if J and K belong to $\mathcal{K}(\mathbb{R}^m)$, then the Hausdorff distance between J and K is

$$d_\text{H}(J,K) = \max\{d^{(m)}(J,K), d^{(m)}(K,J)\}.$$

We show that d_H is a metric on $\mathcal{K}(\mathbb{R}^m)$ by showing that d_H has the metric properties (M1)–(M3).

(M1) Let $J, K \in \mathcal{K}(\mathbb{R}^m)$. By the definition of d_H it follows that $d_\text{H}(J,K) \geq 0$ and $d_\text{H}(K,K) = 0$.

It remains only to show that if $d_\text{H}(J,K) = 0$, then $J = K$.

If $d_\text{H}(J,K) = 0$, then both $d^{(m)}(J,K) = 0$ and $d^{(m)}(K,J) = 0$. Thus,

$$d^{(m)}(\mathbf{j}, K) = 0 \quad \text{for each } \mathbf{j} \in J,$$

and hence $\mathbf{j} \in K$, by Lemma 2.1. So $J \subseteq K$.

Similarly, $K \subseteq J$, and so $J = K$ as required.

Thus (M1) is satisfied.

(M2) Let $J, K \in \mathcal{K}(\mathbb{R}^m)$. By the definition,

$$d_\text{H}(J,K) = d_\text{H}(K,J),$$

and so (M2) is satisfied.

(M3) Let $J, K, L \in \mathcal{K}(\mathbb{R}^m)$. By Lemma 2.3,

$$d^{(m)}(J,L) \leq d^{(m)}(J,K) + d^{(m)}(K,L)$$

and

$$d^{(m)}(L,J) \leq d^{(m)}(L,K) + d^{(m)}(K,J).$$

Since $\max\{a+b, c+d\} \leq \max\{a,d\} + \max\{b,c\}$, for real numbers a, b, c, d, we obtain

$$\begin{aligned}
d_\text{H}(J,L) &= \max\{d^{(m)}(J,L), d^{(m)}(L,J)\} \\
&\leq \max\{d^{(m)}(J,K) + d^{(m)}(K,L), d^{(m)}(L,K) + d^{(m)}(K,J)\} \\
&\leq \max\{d^{(m)}(J,K), d^{(m)}(K,J)\} + \max\{d^{(m)}(K,L), d^{(m)}(L,K)\} \\
&= d_\text{H}(J,K) + d_\text{H}(K,L).
\end{aligned}$$

Hence (M3) is satisfied.

Since (M1)–(M3) are satisfied, d_H is a metric on $\mathcal{K}(\mathbb{R}^m)$. ∎

We now have a metric space $(\mathcal{K}(\mathbb{R}^m), d_\text{H})$. Before we can apply the Contraction Mapping Theorem, we need the following result.

Theorem 2.5

The metric space $(\mathcal{K}(\mathbb{R}^m), d_\text{H})$ is complete.

Recall that a metric space is complete if each Cauchy sequence is convergent. The proof of this theorem is long and complicated. We give it in Subsection 2.4 — for completeness!

2.2 Constructing contraction mappings

We now have a complete metric space $(\mathcal{K}(\mathbb{R}^m), d_H)$ to which we can apply the Contraction Mapping Theorem. Our aim is to see how the methods we used in Section 1 to obtain examples of fractals can be described in terms of mappings on $\mathcal{K}(\mathbb{R}^m)$.

In Section 1, we took families of similarities $S_1, S_2, \ldots, S_k$ defined on $\mathbb{R}$ or $\mathbb{R}^2$, and repeatedly applied them to a given compact set K_0. So if

$$K_1 = \bigcup_{i=1}^{k} S_i(K_0)$$

and

$$K_n = \bigcup_{i=1}^{k} S_i(K_{n-1}) \quad \text{for } n \geq 2,$$

then our fractal set is the 'limit' of the sequence of compact sets (K_n). Now that we have a metric for the collections of compact subsets of $\mathbb{R}$ and $\mathbb{R}^2$, we know what the limit of a sequence of compact sets is — it is a compact set J for which

$$d_H(K_n, J) \to 0 \text{ as } n \to \infty.$$

It remains for us to show that *there is* a limit for a sequence of sets K_n determined by similarities $S_1, \ldots, S_k$, and that this limit does not depend on our choice of starting set K_0. We use the Contraction Mapping Theorem for this.

In order to apply the Contraction Mapping Theorem, we describe the construction process in terms of a single mapping from $\mathcal{K}(\mathbb{R})$ to $\mathcal{K}(\mathbb{R})$. So let us define the mapping $S: \mathcal{K}(\mathbb{R}) \to \mathcal{K}(\mathbb{R})$ by

$$S(K) = S_1(K) \cup S_2(K) \cup \cdots \cup S_k(K) \quad \text{for each } K \in \mathcal{K}(\mathbb{R}).$$

Then

$$K_n = S(K_{n-1}) \quad \text{for each } n \in \mathbb{N}.$$

Problem 2.7

Let $S_1, S_2: \mathbb{R} \to \mathbb{R}$ be defined by

$$S_1(x) = \tfrac{1}{3}x \quad \text{and} \quad S_2(x) = \tfrac{1}{3}(x+2).$$

Let S be the mapping on $\mathcal{K}(\mathbb{R})$ defined by

$$S(K) = S_1(K) \cup S_2(K),$$

for each $K \in \mathcal{K}(\mathbb{R})$. Find the image of each of the following sets under S, and state whether the image belongs to $\mathcal{K}(\mathbb{R})$.

(a) $[0, 2]$ (b) $[-2, 2]$

In the above problem, you should have found that each set is mapped by S to a set in $\mathcal{K}(\mathbb{R})$. In fact, *any* set in $\mathcal{K}(\mathbb{R})$ is mapped by S to another set in $\mathcal{K}(\mathbb{R})$, since both S_1 and S_2 are continuous and so map compact sets to compact sets. Further, the union of two compact sets is also compact, and so S is indeed a mapping from $\mathcal{K}(\mathbb{R})$ to itself.

Unit C2, Theorem 2.2.
See the Exercises for Unit C2.

Similar arguments can be used to prove the following more general result.

Proposition 2.6

Let $S_1, S_2, \ldots, S_k \colon \mathbb{R}^m \to \mathbb{R}^m$ be $(d^{(m)}, d^{(m)})$-continuous, and let S be defined by
$$S(K) = \bigcup_{i=1}^{k} S_i(K) \quad \text{for each } K \in \mathcal{K}(\mathbb{R}^m).$$
Then S maps $\mathcal{K}(\mathbb{R}^m)$ to $\mathcal{K}(\mathbb{R}^m)$.

Remark

Note that each of the fractals in Section 1 is obtained by repeatedly applying mappings of the form described in Proposition 2.6.

Finally, in order to apply the Contraction Mapping Theorem, we require a mapping from $\mathcal{K}(\mathbb{R}^m)$ to itself that is a *contraction mapping*. Recall that a contraction mapping is defined as follows.

See *Unit C4*, Section 3.

Definition

Let (X, d) be a metric space. A function $T \colon X \to X$ is a **contraction mapping** if there is a real number $0 \leq \lambda < 1$ such that
$$d(T(x), T(y)) \leq \lambda d(x, y) \quad \text{for all } x, y \in X.$$
Any such real number λ is a **contraction ratio** for T.

Remark

Note that any similarity with similarity ratio $c < 1$ is a contraction mapping on $\mathbb{R}^m$ with contraction ratio c. In particular, all the similarities that are used to define fractals in Section 1 are contraction mappings.

In general, suppose that $S_1, S_2, \ldots, S_k \colon \mathbb{R}^m \to \mathbb{R}^m$ are continuous functions and consider the mapping $S \colon \mathcal{K}(\mathbb{R}^m) \to \mathcal{K}(\mathbb{R}^m)$ defined by
$$S(K) = \bigcup_{i=1}^{k} S_i(K) \quad \text{for each } K \in \mathcal{K}(\mathbb{R}^m).$$
We now show that if each of the functions $S_1, S_2, \ldots, S_k$ is a contraction mapping on $\mathbb{R}^m$, then the function S is also a contraction mapping on $\mathcal{K}(\mathbb{R}^m)$.

Proposition 2.7

Let $S_1, S_2, \ldots, S_k \colon \mathbb{R}^m \to \mathbb{R}^m$ be contraction mappings. Then the mapping $S \colon \mathcal{K}(\mathbb{R}^m) \to \mathcal{K}(\mathbb{R}^m)$ defined by
$$S(K) = \bigcup_{i=1}^{k} S_i(K) \quad \text{for each } K \in \mathcal{K}(\mathbb{R}^m),$$
is a contraction mapping with respect to the Hausdorff distance d_H on $\mathcal{K}(\mathbb{R}^m)$; that is, there is a number λ such that $0 \leq \lambda < 1$ and
$$d_H(S(J), S(K)) \leq \lambda d_H(J, K),$$
for all $J, K \in \mathcal{K}(\mathbb{R}^m)$.

Proof Each of the mappings $S_1, S_2, \ldots, S_k$ is a contraction mapping, and so there are $0 \leq \lambda_1, \lambda_2, \ldots, \lambda_k < 1$ such that

$$d^{(m)}(S_i(\mathbf{x}), S_i(\mathbf{y})) \leq \lambda_i d^{(m)}(\mathbf{x}, \mathbf{y}), \quad 1 \leq i \leq k,$$

for all $\mathbf{x}, \mathbf{y} \in \mathbb{R}^m$.

Now let $J, K \in \mathcal{K}(\mathbb{R}^m)$. We show that, for $1 \leq i \leq k$,

$$d^{(m)}(S_i(J), S_i(K)) \leq \lambda_i d^{(m)}(J, K). \tag{2.1}$$

Indeed, for each $\mathbf{j} \in J$,

$$d^{(m)}(S_i(\mathbf{j}), S_i(K)) = \min_{\mathbf{k} \in K} d^{(m)}(S_i(\mathbf{j}), S_i(\mathbf{k}))$$
$$\leq \min_{\mathbf{k} \in K} \lambda_i d^{(m)}(\mathbf{j}, \mathbf{k}) = \lambda_i d^{(m)}(\mathbf{j}, K).$$

So

$$d^{(m)}(S_i(J), S_i(K)) = \max_{\mathbf{j} \in J} d^{(m)}(S_i(\mathbf{j}), S_i(K))$$
$$\leq \max_{\mathbf{j} \in J} \lambda_i d^{(m)}(\mathbf{j}, K) = \lambda_i d^{(m)}(J, K).$$

Thus (2.1) is true.

Now

$$d^{(m)}(S(J), S(K)) = d^{(m)}\left(\bigcup_{i=1}^{k} S_i(J), \bigcup_{j=1}^{k} S_j(K)\right)$$
$$= \max_{\mathbf{x} \in \bigcup_{i=1}^{k} S_i(J)} d^{(m)}\left(\mathbf{x}, \bigcup_{j=1}^{k} S_j(K)\right)$$
$$= \max_{1 \leq i \leq k} \max_{\mathbf{x} \in S_i(J)} d^{(m)}\left(\mathbf{x}, \bigcup_{j=1}^{k} S_j(K)\right)$$
$$= \max_{1 \leq i \leq k} d^{(m)}\left(S_i(J), \bigcup_{j=1}^{k} S_j(K)\right).$$

Since $S_i(K) \subseteq \bigcup_{j=1}^{k} S_j(K)$ for $i = 1, 2, \ldots, k$, the result of Problem 2.5(b) implies that

$$d^{(m)}(S(J), S(K)) \leq \max_{1 \leq i \leq k} d^{(m)}(S_i(J), S_i(K)).$$

This estimate, together with inequality (2.1), implies that

$$d^{(m)}(S(J), S(K)) \leq \max_{1 \leq i \leq k} \lambda_i d^{(m)}(J, K) = \left(\max_{1 \leq i \leq k} \lambda_i\right) d^{(m)}(J, K).$$

The same arguments show that

$$d^{(m)}(S(K), S(J)) \leq \max_{1 \leq i \leq k} \lambda_i d^{(m)}(K, J) = \left(\max_{1 \leq i \leq k} \lambda_i\right) d^{(m)}(K, J),$$

and so

$$d_{\mathrm{H}}(S(J), S(K)) \leq \left(\max_{1 \leq i \leq k} \lambda_i\right) d_{\mathrm{H}}(J, K).$$

But $0 \leq \max_{1 \leq i \leq k} \lambda_i < 1$, and so S is a contraction mapping on $\mathcal{K}(\mathbb{R}^m)$ with contraction ratio $\lambda = \max_{1 \leq i \leq k} \lambda_i$. ∎

You may omit this proof on a first reading.

2.3 Iterated function schemes

We have shown that, for a given family of contraction mappings $S_1, S_2, \ldots, S_k \colon \mathbb{R}^m \to \mathbb{R}^m$, the mapping $S \colon \mathcal{K}(\mathbb{R}^m) \to \mathcal{K}(\mathbb{R}^m)$ defined by

$$S(K) = \bigcup_{i=1}^{k} S_i(K) \quad \text{for each } K \in \mathcal{K}(\mathbb{R}^m),$$

is a contraction mapping with respect to the Hausdorff distance d_H on $\mathcal{K}(\mathbb{R}^m)$. Since $(\mathcal{K}(\mathbb{R}^m), d_\mathrm{H})$ is a complete metric space, we can apply the Contraction Mapping Theorem to obtain the following important result.

Theorem 2.8

Let $S_1, S_2, \ldots, S_k \colon \mathbb{R}^m \to \mathbb{R}^m$ be contraction mappings, and let the mapping $S \colon \mathcal{K}(\mathbb{R}^m) \to \mathcal{K}(\mathbb{R}^m)$ be defined by

$$S(K) = \bigcup_{i=1}^{k} S_i(K) \quad \text{for each } K \in \mathcal{K}(\mathbb{R}^m).$$

Then there is a unique set $K_S \in \mathcal{K}(\mathbb{R}^m)$ such that

$$S(K_S) = K_S.$$

Moreover, for each $K \in \mathcal{K}(\mathbb{R}^m)$, the sequence of iterates

$$K,\ S(K),\ S^2(K),\ S^3(K),\ \ldots$$

converges to K_S; that is,

$$d_\mathrm{H}(S^n(K), K_S) \to 0 \text{ as } n \to \infty.$$

This is our main result.

Remark

This result is important because for many families of contraction mappings $S_1, S_2, \ldots, S_k$, the set K_S is a fractal. In fact, this result gives a simple method of constructing these fractals — we take a compact set and repeatedly apply the mapping S. This method of generating a fractal is called an *iterated function scheme*.

Since

$$S(K_S) = \bigcup_{i=1}^{k} S_i(K_S) = K_S,$$

we say that the set K_S is *invariant* for the contraction mappings $S_1, S_2, \ldots, S_k$.

Definition

A set A is **invariant** for the mappings $S_1, S_2, \ldots, S_k \colon \mathbb{R}^m \to \mathbb{R}^m$ if

$$\bigcup_{i=1}^{k} S_i(A) = A.$$

Remark

A self-similar set is just a set that is invariant for *similarities* $S_1, S_2, \ldots, S_k \colon \mathbb{R}^m \to \mathbb{R}^m$.

We now have a rigorous justification of the process that we used in Section 1 to generate fractals.

The von Koch curve

To construct the von Koch curve, we take the compact set $K_0 = [0, 1]$ and repeatedly apply the mapping $S: \mathcal{K}(\mathbb{R}^2) \to \mathcal{K}(\mathbb{R}^2)$ given by

$$S(K) = S_1(K) \cup S_2(K) \cup S_3(K) \cup S_4(K) \quad \text{for each } K \in \mathcal{K}(\mathbb{R}^2),$$

where the similarities $S_1, S_2, S_3, S_4: \mathbb{R}^2 \to \mathbb{R}^2$ are as follows.

S_1 scales by a factor of $\frac{1}{3}$.

S_2 scales by a factor of $\frac{1}{3}$, rotates anticlockwise through $\frac{1}{3}\pi$ about the point $(0, 0)$, and translates by $(\frac{1}{3}, 0)$.

S_3 scales by a factor of $\frac{1}{3}$, rotates clockwise by $\frac{1}{3}\pi$ about the point $(0, 0)$, and translates by $(\frac{1}{2}, \frac{\sqrt{3}}{6})$.

S_4 scales by a factor of $\frac{1}{3}$ and translates by $(\frac{2}{3}, 0)$.

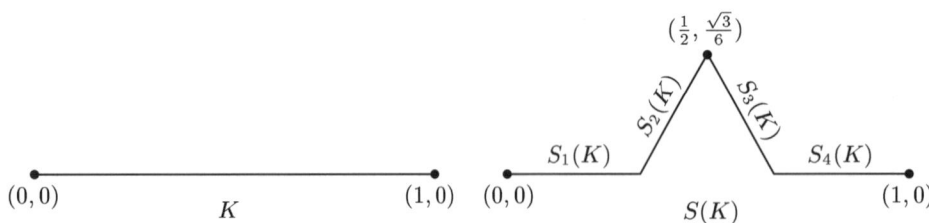

Figure 2.6

We define the von Koch curve to be the limit of the curves $K_n = S^n(K_0)$. In Section 1, we did not justify the existence of such a limit. Since each of the mappings S_1, S_2, S_3, S_4 is a contraction mapping (each with contraction ratio $\frac{1}{3}$), it now follows from Theorem 2.8 that such a limit does indeed exist.

We also claimed in Section 1 that the von Koch curve K_V satisfies the following equation, and is therefore self-similar:

$$K_V = S_1(K_V) \cup S_2(K_V) \cup S_3(K_V) \cup S_4(K_V) = S(K_V),$$

that is, K_V is invariant for S_1, S_2, S_3, S_4. This also follows from Theorem 2.8.

Many other fractal curves can be constructed similarly, as Problem 2.8 illustrates.

Problem 2.8

Let $S_1, S_2, S_3, S_4, S_5: \mathbb{R}^2 \to \mathbb{R}^2$ be similarities defined by

$$S_1(x, y) = \left(\tfrac{1}{3}x, \tfrac{1}{3}y\right), \qquad S_2(x, y) = \left(\tfrac{1}{3}(y + 1), \tfrac{1}{3}x\right),$$
$$S_3(x, y) = \left(\tfrac{1}{3}(x + 1), \tfrac{1}{3}(y + 1)\right), \quad S_4(x, y) = \left(\tfrac{1}{3}(-y + 2), \tfrac{1}{3}x\right),$$
$$S_5(x, y) = \left(\tfrac{1}{3}(x + 2), \tfrac{1}{3}y\right).$$

Let

$$S(K) = \bigcup_{i=1}^{5} S_i(K) \quad \text{for each } K \in \mathcal{K}(\mathbb{R}^2),$$

and let $K_0 = \{(x, y) : 0 \leq x \leq 1, y = 0\}$.

(a) Sketch the sets $S(K_0)$ and $S^2(K_0)$.

(b) Determine the length of $S^n(K_0)$, for each $n \in \mathbb{N}$.

(c) Determine whether $\lim_{n \to \infty} S^n(K_0)$ exists.

Cantor sets

In Section 1, our first example of a fractal was the middle-third Cantor set. To construct it, we take the compact set $K_0 = [0,1]$ and repeatedly apply the mapping $S : \mathcal{K}(\mathbb{R}) \to \mathcal{K}(\mathbb{R})$ defined by

$$S(K) = S_1(K) \cup S_2(K),$$

where

$$S_1(x) = \tfrac{1}{3}x \quad \text{and} \quad S_2(x) = \tfrac{1}{3}(x+2).$$

We define the middle-third Cantor set to be the *intersection* of the nested sequence of sets $K_n = S^n(K_0)$. Since each of the mappings S_1 and S_2 is a contraction mapping (with contraction ratio $\tfrac{1}{3}$), it follows from Theorem 2.8 that the sets K_n converge to a unique limit set K_S (with respect to Hausdorff distance) as $n \to \infty$. It seems reasonable to expect that this limit set is the same as K_C, the middle-third Cantor set. The following lemma shows this to be the case.

Lemma 2.9

Let $S_1, S_2, \ldots, S_k : \mathbb{R}^m \to \mathbb{R}^m$ be contraction mappings, and let $S : \mathcal{K}(\mathbb{R}^m) \to \mathcal{K}(\mathbb{R}^m)$ be the mapping defined by

$$S(K) = \bigcup_{i=1}^{k} S_i(K) \quad \text{for each } K \in \mathcal{K}(\mathbb{R}^m).$$

Let $S(K_0) \subseteq K_0$, for some $K_0 \in \mathcal{K}(\mathbb{R}^m)$; then

$$\bigcap_{n \in \mathbb{N}} S^n(K_0) = K_S,$$

where K_S is the unique set in $\mathcal{K}(\mathbb{R}^m)$ such that $S(K_S) = K_S$.

Proof For simplicity, let $K_n = S^n(K_0)$ and $J = \bigcap_{n \in \mathbb{N}} K_n$. Then, for each $n \in \mathbb{N}$,

$$K_{n+1} = S^{n+1}(K_0) = S^n(S(K_0)) \subseteq S^n(K_0) = K_n,$$

and so the sets K_n are nested.

To show that $J = K_S$, we show that both $J - K_S$ and $K_S - J$ are empty, and so $J \subseteq K_S$ and $K_S \subseteq J$.

First, suppose that $J - K_S \neq \emptyset$, and so there is an $\mathbf{x} \in J - K_S$. *We use a proof by contradiction.*

Then by Lemma 2.1 there is an $\varepsilon > 0$ such that

$$d^{(m)}(\mathbf{x}, K_S) = \varepsilon.$$

Since $\mathbf{x} \in J$, it follows that $\mathbf{x} \in K_n$, for each $n \in \mathbb{N}$, and so

$$d_{\mathrm{H}}(K_n, K_S) \geq d^{(m)}(K_n, K_S) \geq d^{(m)}(\mathbf{x}, K_S) = \varepsilon.$$

This contradicts Theorem 2.8, and so there is no $\mathbf{x} \in J - K_S$; thus $J \subseteq K_S$.

Now suppose that $K_S - J \neq \emptyset$, and so there is an $\mathbf{x} \in K_S - J$.

Since $\mathbf{x} \notin J$, there is an $N \in \mathbb{N}$ with $\mathbf{x} \notin K_N$. Hence by Lemma 2.1 there is an $\varepsilon > 0$ such that

$$d^{(m)}(\mathbf{x}, K_N) = \varepsilon.$$

Since the sets K_n are nested, we find that, for each $n \geq N$,

$$d_H(K_n, K_S) \geq d^{(m)}(K_S, K_n) \geq d^{(m)}(\mathbf{x}, K_n)$$
$$\geq d^{(m)}(\mathbf{x}, K_N) \quad \text{by Lemma 2.2,}$$
$$= \varepsilon.$$

This contradicts Theorem 2.8, and so there is no $\mathbf{x} \in K_S - J$; thus $K_S \subseteq J$. Hence $K_S = J$, as required. ∎

So, as we expected, the middle-third Cantor set $K_C = \bigcap_{n \in \mathbb{N}} K_n$ is the limit of the sets $K_n = S^n(K_0)$. It follows from Theorem 2.8 that the middle-third Cantor set K_C satisfies the following equation, and is therefore self-similar:

$$K_C = S_1(K_C) \cup S_2(K_C) = S(K_C). \tag{2.2}$$

In fact, Theorem 2.8 tells us that the middle-third Cantor set is the *only* set in $\mathcal{K}(\mathbb{R})$ that is invariant for S_1 and S_2. This means that the middle-third Cantor set can be *defined* as the unique non-empty compact solution to (2.2). So this intricate set can be defined by a simple equation.

Note that there are solutions to (2.2) that are *not* non-empty and compact, as the following problem illustrates.

Problem 2.9

Show that for K equal to any of the sets $\varnothing$, $\mathbb{R}$ and $[0, \infty)$,

$$K = S_1(K) \cup S_2(K),$$

where $S_1, S_2 \colon \mathbb{R} \to \mathbb{R}$ are given by $S_1(x) = \frac{1}{3}x$ and $S_2(x) = \frac{1}{3}(x + 2)$.

It follows from Theorem 2.8 that if we take K_0 to be *any* non-empty compact subset of $\mathbb{R}$, then the middle-third Cantor set is the limit of the sets $K_n = S^n(K_0)$. Further, it follows from Lemma 2.9 that if $S(K_0) \subseteq K_0$, then the middle-third Cantor set is the intersection of the sets K_n.

Problem 2.10

Let K_C be the middle-third Cantor set, and let $S \colon \mathcal{K}(\mathbb{R}) \to \mathcal{K}(\mathbb{R})$ be the mapping defined by

$$S(K) = S_1(K) \cup S_2(K),$$

where $S_1, S_2 \colon \mathbb{R} \to \mathbb{R}$ are given by

$$S_1(x) = \tfrac{1}{3}x \quad \text{and} \quad S_2(x) = \tfrac{1}{3}(x + 2).$$

Show that

$$\bigcap_{k=0}^{\infty} S^k([0, 2]) = K_C.$$

Note that if K_0 is *not* compact, then the sets $K_n = S^n(K_0)$ do not necessarily converge to the middle-third Cantor set. For example, if $K_0 = [0, \infty)$, then

$$S(K_0) = S_1(K_0) \cup S_2(K_0) = [0, \infty) \cup [\tfrac{2}{3}, \infty) = [0, \infty) = K_0,$$

and so $S^n(K_0) = K_0$, for each $n \in \mathbb{N}$.

The Sierpiński gasket

Our other main example of a fractal is the Sierpiński gasket. To construct it, we take K_0 to be the equilateral triangle whose base is the line segment from $(0,0)$ to $(1,0)$, and repeatedly apply the mapping $S: \mathcal{K}(\mathbb{R}^2) \to \mathcal{K}(\mathbb{R}^2)$ given by

$$S(K) = S_1(K) \cup S_2(K) \cup S_3(K),$$

where the mappings $S_1, S_2, S_3: \mathbb{R}^2 \to \mathbb{R}^2$ are defined by

$$S_1(x,y) = \left(\tfrac{1}{2}x, \tfrac{1}{2}y\right), \qquad S_2(x,y) = \left(\tfrac{1}{2}(x+1), \tfrac{1}{2}y\right),$$
$$S_3(x,y) = \left(\tfrac{1}{4}(2x+1), \tfrac{1}{4}(2y+\sqrt{3})\right).$$

We define the Sierpiński gasket to be the *intersection* of the nested sequence of sets $K_n = S^n(K_0)$. Since each of the mappings S_1, S_2 and S_3 is a contraction mapping (with contraction ratio $\tfrac{1}{2}$), it follows from Theorem 2.8 that the sets K_n converge to a unique limit set (with respect to Hausdorff distance) as $n \to \infty$. Since $S(K_0) \subset K_0$, it follows from Lemma 2.9 that this limit set is the Sierpiński gasket.

We now know from Theorem 2.8 that the Sierpiński gasket can be obtained by repeatedly applying the mapping S to *any* non-empty compact subset of $\mathbb{R}^2$. Further, it follows from Theorem 2.8 that the Sierpiński gasket can be defined to be the unique non-empty compact set that is invariant for S_1, S_2 and S_3.

In the next problem, we ask you to show how similarities can also be used to define simple sets.

Problem 2.11

Let $S: \mathcal{K}(\mathbb{R}^2) \to \mathcal{K}(\mathbb{R}^2)$ be defined by

$$S(K) = S_1(K) \cup S_2(K),$$

where $S_1, S_2: \mathbb{R}^2 \to \mathbb{R}^2$ are given by

$$S_1(x,y) = \left(\tfrac{1}{2}x, \tfrac{1}{2}y\right) \quad \text{and} \quad S_2(x,y) = \left(\tfrac{1}{2}x + \tfrac{1}{2}, \tfrac{1}{2}y + \tfrac{1}{2}\right).$$

(a) Sketch the sets $K_0 = \{(x,y) : 0 \leq x, y \leq 1\}$ and $K_1 = S(K_0)$.

(b) Let $K_n = S^n(K_0)$, for each $n \in \mathbb{N}$. Verify that

$$\lim_{n \to \infty} K_n = \{(x,y) : y = x, 0 \leq x \leq 1\}.$$

2.4 Proof of Theorem 2.5

We now prove that $(\mathcal{K}(\mathbb{R}^m), d_H)$ is a complete metric space. Recall that a metric space (X, d) is complete if each Cauchy sequence in X converges to a point $x \in X$.

This proof will not be assessed.

See Unit C4, Section 1.

Let (K_n) be a Cauchy sequence in $\mathcal{K}(\mathbb{R}^m)$. Then K_n is a non-empty compact set for each $n \in \mathbb{N}$, and for each $\varepsilon > 0$, there is an $N \in \mathbb{N}$ such that

$$d_H(K_n, K_{n'}) < \varepsilon \quad \text{for all } n, n' > N.$$

In order to show that $(\mathcal{K}(\mathbb{R}^m), d_H)$ is complete, we find a non-empty compact set J with $d_H(K_n, J) \to 0$ as $n \to \infty$. Since (K_n) is a Cauchy sequence, it is enough to find a convergent subsequence of (K_n).

See Unit C4, Lemma 1.4.

Since (K_n) is a Cauchy sequence, for each $i \in \mathbb{N} \cup \{0\}$ there is an $N(i) \in \mathbb{N}$ with $N(i+1) > N(i)$ such that

$$d_H(K_n, K_{n'}) < \left(\tfrac{1}{2}\right)^i \quad \text{for all } n, n' > N(i). \tag{2.3}$$

We show that $(K_{N(i)})_{i \in \mathbb{N}}$ is d_H-convergent in $\mathcal{K}(\mathbb{R}^m)$.

We start by finding a candidate for the limit of $K_{N(i)}$ as $i \to \infty$.

We consider the sets

$$J_i = \left\{ \mathbf{x} : d^{(m)}(\mathbf{x}, K_{N(i)}) \leq \left(\tfrac{1}{2}\right)^{i-2} \right\} \quad \text{for } i \in \mathbb{N}.$$

If $\mathbf{x} \in K_{N(i)}$, then it follows from Lemma 2.1 that $d^{(m)}(\mathbf{x}, K_{N(i)}) = 0$, and so $\mathbf{x} \in J_i$. Thus, $K_{N(i)} \subseteq J_i$ and hence, by Problem 2.5(a),

$$d^{(m)}(K_{N(i)}, J_i) = 0. \tag{2.4}$$

Notice also that

$$d^{(m)}(J_i, K_{N(i)}) = \max\{ d^{(m)}(\mathbf{x}, K_{N(i)}) : \mathbf{x} \in J_i \} \leq \left(\tfrac{1}{2}\right)^{i-2}, \tag{2.5}$$

and so

$$d_H(J_i, K_{N(i)}) \leq \left(\tfrac{1}{2}\right)^{i-2}. \tag{2.6}$$

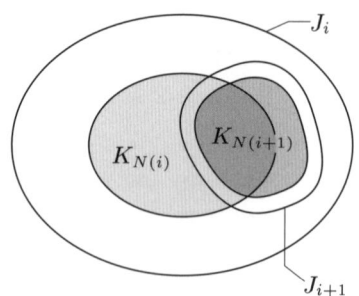

Figure 2.7

Let

$$J = \bigcap_{i \in \mathbb{N}} J_i.$$

This set is our candidate for the limit of the sequence $(K_{N(i)})_{i \in \mathbb{N}}$.

We show that

(a) $J \in \mathcal{K}(\mathbb{R}^m)$;

(b) $\lim_{i \to \infty} d_H(K_{N(i)}, J) = 0$.

Proof of (a)

We start by noting that each of the sets J_i is a non-empty compact set and so belongs to $\mathcal{K}(\mathbb{R}^m)$.

We ask you to prove this in the Exercises.

We now estimate the $d^{(m)}$-distance of J_{i+1} from $K_{N(i)}$. It follows from Lemma 2.3, (2.5) and (2.3) that

$$\begin{aligned} d^{(m)}(J_{i+1}, K_{N(i)}) &\leq d^{(m)}(J_{i+1}, K_{N(i+1)}) + d^{(m)}(K_{N(i+1)}, K_{N(i)}) \\ &\leq \left(\tfrac{1}{2}\right)^{i-1} + d^{(m)}(K_{N(i+1)}, K_{N(i)}) \\ &< \left(\tfrac{1}{2}\right)^{i-1} + \left(\tfrac{1}{2}\right)^{i-1} = \left(\tfrac{1}{2}\right)^{i-2}. \end{aligned}$$

When applying inequality (2.3), we use the fact that $N(i+1) > N(i) > N(i-1)$.

Hence if $\mathbf{x} \in J_{i+1}$, then $d^{(m)}(\mathbf{x}, K_{N(i)}) < 1/2^{i-2}$; this implies that $\mathbf{x} \in J_i$, and so $J_{i+1} \subseteq J_i$ for each $i \in \mathbb{N}$. Thus J is the intersection of a nested sequence of non-empty compact sets and is therefore non-empty and compact.

Unit C2, Theorem 3.10.

Proof of (b)

To prove (b) it is enough to show that

$$d_{\mathrm{H}}(J_i, J) \to 0 \text{ as } i \to \infty, \qquad (2.7)$$

since (M3) for d_{H} together with (2.6) implies that

$$d_{\mathrm{H}}(K_{N(i)}, J) \leq d_{\mathrm{H}}(K_{N(i)}, J_i) + d_{\mathrm{H}}(J_i, J) \leq \left(\tfrac{1}{2}\right)^{i-2} + d_{\mathrm{H}}(J_i, J),$$

and so if $d_{\mathrm{H}}(J_i, J) \to 0$ as $i \to \infty$, then $d_{\mathrm{H}}(K_{N(i)}, J) \to 0$ as $i \to \infty$.

To show that (2.7) holds, we need show only that $d^{(m)}(J_i, J) \to 0$ as $i \to \infty$, since $J \subseteq J_i$ for each i implies that $d^{(m)}(J, J_i) = 0$ (by Problem 2.5(a)).

Thus for each $i \in \mathbb{N}$, we need to find an upper estimate for

$$d^{(m)}(J_i, J) = \max\{d^{(m)}(\mathbf{x}, J) : \mathbf{x} \in J_i\}.$$

So we fix $i \in \mathbb{N}$ and let $\mathbf{x} \in J_i$.

We define a sequence of points $(\mathbf{y}_j)_{j \in \mathbb{N}}$ as follows. Take

$\mathbf{y}_1 \in J_{i+1}$ such that $d^{(m)}(\mathbf{x}, \mathbf{y}_1) \leq \left(\tfrac{1}{2}\right)^{i-2}$,

$\mathbf{y}_2 \in J_{i+2}$ such that $d^{(m)}(\mathbf{y}_1, \mathbf{y}_2) \leq \left(\tfrac{1}{2}\right)^{i-1}$,

and, more generally, for $j \geq 2$ choose

$$\mathbf{y}_j \in J_{i+j} \text{ such that } d^{(m)}(\mathbf{y}_{j-1}, \mathbf{y}_j) \leq \left(\tfrac{1}{2}\right)^{i+j-3}. \qquad (2.8)$$

This is possible since it follows from Lemma 2.3, (2.5), (2.3) and (2.4) that

$$\begin{aligned} d^{(m)}(J_{i+j}, J_{i+j-1}) &\leq d^{(m)}(J_{i+j}, K_{N(i+j)}) + d^{(m)}(K_{N(i+j)}, K_{N(i+j-1)}) \\ &\quad + d^{(m)}(K_{N(i+j-1)}, J_{i+j-1}) \\ &< \left(\tfrac{1}{2}\right)^{i+j-2} + \left(\tfrac{1}{2}\right)^{i+j-2} + 0 = \left(\tfrac{1}{2}\right)^{i+j-3}. \end{aligned}$$

It is easy to check that $(\mathbf{y}_j)$ is a $d^{(m)}$-Cauchy sequence in $\mathbb{R}^m$: for if $n, n' \in \mathbb{N}$ with $n' > n$, then (M3) for $d^{(m)}$ together with (2.8) gives

$$\begin{aligned} d^{(m)}(\mathbf{y}_{n'}, \mathbf{y}_n) &\leq d^{(m)}(\mathbf{y}_{n'}, \mathbf{y}_{n'-1}) + d^{(m)}(\mathbf{y}_{n'-1}, \mathbf{y}_{n'-2}) + \cdots + d^{(m)}(\mathbf{y}_{n+1}, \mathbf{y}_n) \\ &\leq \left(\tfrac{1}{2}\right)^{i+n'-3} + \left(\tfrac{1}{2}\right)^{i+n'-4} + \cdots + \left(\tfrac{1}{2}\right)^{i+n-2} \\ &\leq \left(\tfrac{1}{2}\right)^{i+n-1} \to 0 \quad \text{as } n \to \infty. \end{aligned}$$

Since $(\mathbb{R}^m, d^{(m)})$ is complete, the sequence has a limit $\mathbf{y} \in \mathbb{R}^m$.

Moreover, since the sequence $(\mathbf{y}_j)$ lies in the compact set J_i, its limit $\mathbf{y}$ is also in J_i. For $M \in \mathbb{N}$, the subsequence $(\mathbf{y}_j)_{j \geq M}$ is a Cauchy sequence in the compact set J_{i+M}, and so the limit $\mathbf{y}$ must also be in J_{i+M}. This is true for each $M \in \mathbb{N}$ and so

$$\mathbf{y} \in \bigcap_{M \in \mathbb{N}} J_{i+M} = J.$$

Hence $d^{(m)}(\mathbf{x}, J) \leq d^{(m)}(\mathbf{x}, \mathbf{y})$. By repeatedly using the Triangle Inequality and (2.8), we estimate that, for any $N \in \mathbb{N}$,

$$\begin{aligned}
d^{(m)}(\mathbf{x}, \mathbf{y}) &\leq d^{(m)}(\mathbf{x}, \mathbf{y}_1) + d^{(m)}(\mathbf{y}_1, \mathbf{y}_2) + \cdots + d^{(m)}(\mathbf{y}_{N-1}, \mathbf{y}_N) + d^{(m)}(\mathbf{y}_N, \mathbf{y}) \\
&\leq \left(\tfrac{1}{2}\right)^{i-2} + \left(\tfrac{1}{2}\right)^{i-1} + \cdots + \left(\tfrac{1}{2}\right)^{i+N-3} + d^{(m)}(\mathbf{y}_N, \mathbf{y}) \\
&\leq \sum_{j \in \mathbb{N}} \left(\tfrac{1}{2}\right)^{i+j-3} + d^{(m)}(\mathbf{y}_N, \mathbf{y}) \\
&= \left(\tfrac{1}{2}\right)^{i-3} + d^{(m)}(\mathbf{y}_N, \mathbf{y}).
\end{aligned}$$

However, $d^{(m)}(\mathbf{y}_N, \mathbf{y}) \to 0$ as $N \to \infty$, and so

$$d^{(m)}(\mathbf{x}, \mathbf{y}) \leq \left(\tfrac{1}{2}\right)^{i-3}.$$

This implies that

$$d^{(m)}(\mathbf{x}, J) \leq \left(\tfrac{1}{2}\right)^{i-3}.$$

Since this is true for any point $\mathbf{x} \in J_i$, it follows that

$$d^{(m)}(J_i, J) \leq \left(\tfrac{1}{2}\right)^{i-3},$$

and so $d^{(m)}(J_i, J) \to 0$ as $i \to \infty$, as required.

Since both (a) and (b) hold, $(\mathcal{K}(\mathbb{R}^m), d_{\mathrm{H}})$ is a complete metric space. ∎

3 Dimensions of fractals

After working through this section, you should be able to:
▶ explain how the *box dimension* of a set is calculated;
▶ verify that a given family of similarities satisfies the *open set condition*;
▶ determine the box dimension of a simple fractal.

In Section 1 you saw several examples of fractals, each of which is hard to describe in terms of classical geometry. For example, the von Koch curve seems to be 'smaller' than 2-dimensional sets such as rectangles (since it has zero area) and yet seems to be 'larger' than 1-dimensional sets such as line segments (since it is bounded but does not have finite length).

We now introduce a way of defining dimension that allows the dimension of a set to be a non-integer number. This enables us to assign a dimension to the von Koch curve that is greater than 1 but less than 2.

3.1 Box dimension

There are many definitions of dimension that take non-integer values. Here we look at one of the most widely used — the *box dimension*. We begin with an informal discussion of box dimension; for simplicity, we restrict our attention to subsets of the plane.

Suppose that we wish to estimate the dimension of a bounded set $A \subset \mathbb{R}^2$. We begin by covering the plane with a grid made up of boxes of side d. We then count how many boxes in the grid intersect the set A.

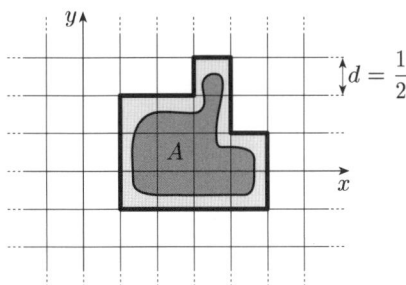

Figure 3.1

We adopt the following notation.

Definition

Let $d \in (0, 1)$. The **grid** G_d is the set of all boxes of side d:
$$G_d = \{[nd, (n+1)d] \times [md, (m+1)d] : n, m \in \mathbb{Z}\}.$$
The number of (closed) boxes in G_d that intersect a set A is denoted by $N_d(A)$.

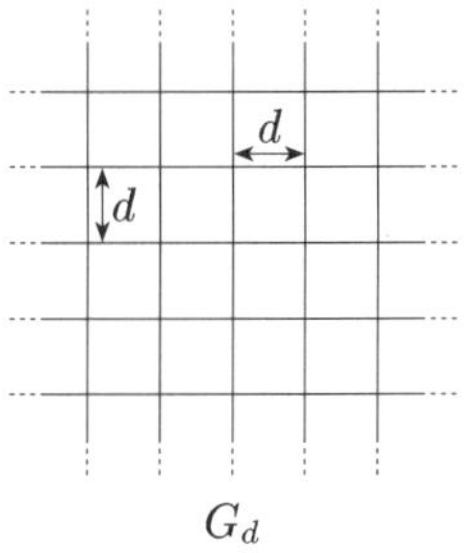

For example, Figure 3.1 shows a set A that meets 12 boxes in the grid $G_{1/2}$, so $N_{1/2}(A) = 12$.

Figure 3.2

Problem 3.1

Let A be a horizontal line segment of length 1. Calculate $N_{1/4}(A)$ when
(a) $A = \{(x,y) : 0 \leq x \leq 1, y = 0\}$;
(b) $A = \{(x,y) : 0 \leq x \leq 1, y = \frac{1}{8}\}$;
(c) $A = \{(x,y) : -\frac{1}{8} \leq x \leq \frac{7}{8}, y = \frac{1}{8}\}$;
(d) $A = \{(x,y) : -\frac{1}{8} \leq x \leq \frac{7}{8}, y = 0\}$.

This problem shows that applying a translation to a set A can change the number $N_d(A)$.

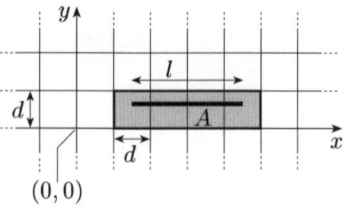

Figure 3.3

Let A be a horizontal line segment of length $l > 0$. For $d > 0$, whatever the position of A relative to the grid G_d, we have

$$N_d(A) \geq \frac{l}{d},$$

since

$$N_d(A) \times \text{(the side-length of each box in } G_d) \geq \text{the length of } A.$$

The number $N_d(A)$ is highest when A is on one of the grid lines and has its endpoints at vertices of the grid (as in Problem 3.1(a)).

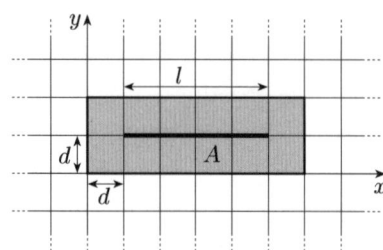

Figure 3.4

In this case, $N_d(A) = 2\left(\frac{l}{d} + 2\right)$. So, if $d < l$, then

$$N_d(A) \leq 2\left(\frac{l}{d} + 2 \times \frac{l}{d}\right) = \frac{6l}{d},$$

so

$$\frac{l}{d} \leq N_d(A) \leq \frac{6l}{d}.$$

Thus, although changing the position of the straight line A relative to the grid can change the value of $N_d(A)$, it can change it by a factor of at most 6. Whatever the position of A, the number $N_d(A)$ 'grows like' $1/d$ for $d < l$.

We now consider the case when A is a square of side $l > 0$ with sides parallel to the axes. Whatever the position of A, we have

$$N_d(A) \geq \left(\frac{l}{d}\right)^2,$$

since

$$N_d(A) \times \text{(the area of each box in } G_d) \geq \text{the area of } A.$$

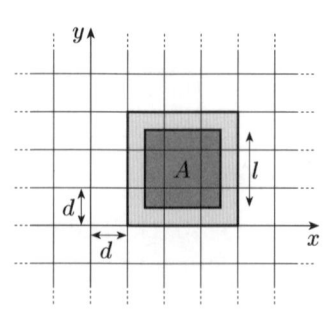

Figure 3.5

The number $N_d(A)$ is highest when the edges of A are on the grid lines and have their endpoints at vertices of the grid.

In this case,
$$N_d(A) = \left(\frac{l}{d} + 2\right) \times \left(\frac{l}{d} + 2\right).$$

So, if $d < l$, then
$$N_d(A) \leq \left(\frac{l}{d} + \left(2 \times \frac{l}{d}\right)\right) \times \left(\frac{l}{d} + \left(2 \times \frac{l}{d}\right)\right) = \frac{3l}{d} \times \frac{3l}{d} = \left(\frac{3l}{d}\right)^2,$$

so
$$\frac{l^2}{d^2} \leq N_d(A) \leq \frac{9l^2}{d^2}.$$

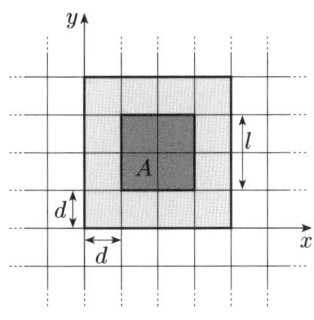

Figure 3.6

Hence, whatever the position of the square A relative to the grid, the number $N_d(A)$ 'grows like' $1/d^2$ for $d < l$.

For both these examples, provided d is sufficiently small, $N_d(A)$ 'grows like' $1/d^s$, where s is our intuitive idea of the dimension of the set A. This suggests that if we have *any* bounded set A for which $N_d(A)$ 'grows like' $1/d^s$, then we can *define* its dimension to be s. This gives a definition of dimension that can take non-integer values, but that agrees with our existing ideas of dimension for simple sets such as line segments and squares.

We now formalize this idea. Suppose that there are numbers m and M such that $0 < m \leq M < \infty$ and
$$\frac{m}{d^s} \leq N_d(A) \leq \frac{M}{d^s} \quad \text{for } 0 < d < 1.$$

Taking logarithms, we obtain
$$\log m - s \log d \leq \log N_d(A) \leq \log M - s \log d.$$

Dividing by $-\log d$, we obtain

These are natural logarithms to base e.

Here $\log d$ is negative, since $0 \leq d < 1$.

$$\frac{\log m}{-\log d} + s \leq \frac{\log N_d(A)}{-\log d} \leq \frac{\log M}{-\log d} + s.$$

Finally, taking the limit as $d \to 0$, we have
$$s = \lim_{d \to 0} \frac{\log N_d(A)}{-\log d}.$$

The limit exists, since m and M do not depend on d.

We now give a formal definition of box dimension.

Definition

Let A be a bounded subset of $\mathbb{R}^2$. The **box dimension** of A is
$$\dim A = \lim_{d \to 0} \frac{\log N_d(A)}{-\log d},$$
provided that this limit exists.

Box dimension is sometimes referred to as *box-counting dimension*.

Worked problem 3.1

Determine the box dimension of the middle-third Cantor set K_C.

Solution

Our objective is to find

$$\lim_{d \to 0} \frac{\log N_d(K_C)}{-\log d}.$$

For a given $d > 0$, we estimate $N_d(K_C)$ by comparing it with $N_d(K_n)$, where n is chosen so that 3^{-n} is approximately equal to d; Figure 3.7 shows $G_{3^{-2}}$ and K_2.

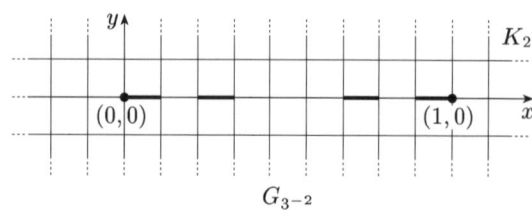

Figure 3.7

We note first that, at the nth stage of the construction of K_C, the set K_n consists of 2^n line segments each of length $\left(\frac{1}{3}\right)^n$. The set K_n contains K_C, and since the endpoints of the intervals making up K_n are not removed in the construction, each of the 2^n intervals of K_n contains a point of K_C.

We regard K_C as a subset of the plane, by embedding it in the x-axis.

Let $d < 1$ and choose an $n \in \mathbb{N}$ so that

$$3^{-n} \le d < 3^{-(n-1)}.$$

We find an upper bound for $N_d(K_C)$ by observing that each of the 2^n line segments making up K_n intersects at most 6 boxes from G_d, and so

$$N_d(K_C) \le N_d(K_n) \le 6 \times 2^n = 3 \times 2^{n+1} \quad \text{since } K_C \subseteq K_n.$$

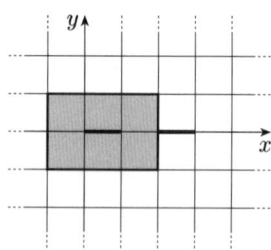

Figure 3.8 *We must count each box with an edge that meets K_n.*

To find a lower bound for $N_d(K_C)$, we note that each interval in K_n contains a point of K_C, and each box in G_d can intersect at most two intervals in K_n, so

$$N_d(K_C) \ge \frac{2^n}{2} = 2^{n-1}.$$

Hence

$$2^{n-1} \le N_d(K_C) \le 3 \times 2^{n+1},$$

and so

$$(n-1)\log 2 \le \log N_d(K_C) \le (n+1)\log 2 + \log 3. \qquad (3.1)$$

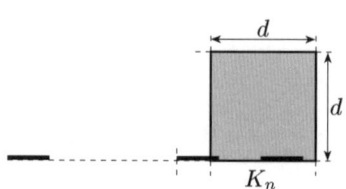

Figure 3.9

Now, since $3^{-n} \le d < 3^{-(n-1)}$,

$$-n \log 3 \le \log d < -(n-1)\log 3,$$

and so, for $n \ge 2$,

$$\frac{1}{n \log 3} \le \frac{1}{-\log d} < \frac{1}{(n-1)\log 3}.$$

Thus for $n \ge 2$ and $3^{-n} \le d \le 3^{-(n-1)}$, we can combine these inequalities with (3.1) to obtain

$$\frac{(n-1)\log 2}{n \log 3} \le \frac{\log N_d(K_C)}{-\log d} < \frac{(n+1)\log 2 + \log 3}{(n-1)\log 3}.$$

Now

$$\lim_{n \to \infty} \frac{(n-1)\log 2}{n \log 3} = \lim_{n \to \infty} \left(\frac{1 - (1/n)}{1}\right)\frac{\log 2}{\log 3} = \frac{\log 2}{\log 3}$$

and
$$\lim_{n\to\infty} \frac{(n+1)\log 2 + \log 3}{(n-1)\log 3} = \lim_{n\to\infty} \left(\frac{1+(1/n)}{1-(1/n)}\right) \frac{\log 2}{\log 3} + \lim_{n\to\infty} \frac{1}{n-1} \frac{\log 2}{\log 3} = \frac{\log 2}{\log 3}.$$

Since $n \to \infty$ as $d \to 0$, we deduce that
$$\dim K_{\mathrm{C}} = \lim_{d\to 0} \frac{\log N_d(K_{\mathrm{C}})}{-\log d} = \frac{\log 2}{\log 3}. \qquad \blacksquare$$

Problem 3.2

Let $S\colon \mathcal{K}(\mathbb{R}) \to \mathcal{K}(\mathbb{R})$ be defined by
$$S(K) = S_1(K) \cup S_2(K),$$
where $S_1(x) = \frac{1}{4}(x+1)$ and $S_2(x) = \frac{1}{4}(x+3)$.

Let $K_0 = [0,1]$ and $K_n = S^n(K_0)$, for each $n \in \mathbb{N}$.

Determine $\dim K_S$, where $K_S = \bigcap_{n\in\mathbb{N}} K_n$.

3.2 The open set condition

We have now found a way of defining the dimension of an intricate set. As you have seen, there may be many calculations involved in determining the dimension of a set. We now wish to calculate the dimensions of the fractals that we constructed in Section 2; each of these is the invariant set of a family of similarities $S_1, S_2, \ldots, S_k$. In this case, there is a simple formula for the dimension of the set, provided that the similarities satisfy a special condition.

> **Definition**
>
> The similarities $S_1, S_2, \ldots, S_k$ satisfy the **open set condition** if there exists a non-empty bounded open set U such that
> $$S_i(U) \cap S_j(U) = \varnothing, \quad \text{for } i \neq j, \quad \text{and} \quad \bigcup_{i=1}^{k} S_i(U) \subseteq U.$$

Remark

The open set condition is a 'disjointness' condition: it says that there is an open set U such that the images of U under distinct similarities are disjoint.

Cantor set

For example, consider the middle-third Cantor set, which is the invariant set of the similarities
$$S_1(x) = \tfrac{1}{3}x \quad \text{and} \quad S_2(x) = \tfrac{1}{3}(x+2).$$

These two similarities satisfy the open set condition, as you can see by taking U to be the open interval $(0,1)$: then $S_1(U) = (0, \tfrac{1}{3})$ and $S_2(U) = (\tfrac{2}{3}, 1)$ so
$$S_1(U) \cap S_2(U) = \varnothing$$
and
$$S_1(U) \cup S_2(U) = (0, \tfrac{1}{3}) \cup (\tfrac{2}{3}, 1) \subset U.$$

Problem 3.3

Let $S_1, S_2, S_3 \colon \mathbb{R}^2 \to \mathbb{R}^2$ be the similarities defined by

$$S_1(x,y) = \left(\tfrac{1}{2}x, \tfrac{1}{2}y\right),$$
$$S_2(x,y) = \left(\tfrac{1}{2}x + \tfrac{1}{2}, \tfrac{1}{2}y\right),$$
$$S_3(x,y) = \left(\tfrac{1}{2}x + \tfrac{1}{4}, \tfrac{1}{2}y + \tfrac{\sqrt{3}}{4}\right).$$

Show that S_1, S_2 and S_3 satisfy the open set condition.

The invariant set of S_1, S_2, S_3 here is the Sierpiński gasket.

We now give the main result of this section.

Theorem 3.1

Let K_S be the invariant set of the similarities $S_1, S_2, \ldots, S_k$, and let $c_i \in (0,1)$ be the similarity ratio of S_i, for $1 \leq i \leq k$. If $S_1, S_2, \ldots, S_k$ satisfy the open set condition, then

$$\dim K_S = s,$$

where s is the solution of the equation $\sum_{i=1}^{k} c_i^s = 1$.

Proof (sketch) The proof of this result is quite complicated, and so we just outline some of the ideas that are involved.

This proof is not assessed.

In particular, we indicate why the box dimension of K_S, if it exists, must be the real number s that satisfies

$$c_1^s + c_2^s + \cdots + c_k^s = 1.$$

To reduce writing, we suppose that we have only two similarities $S_1, S_2 \colon \mathbb{R}^2 \to \mathbb{R}^2$, with similarity ratios c_1 and c_2, respectively.

Since K_S is the invariant set of the similarities, we know that

$$K_S = S_1(K_S) \cup S_2(K_S).$$

The key observation underlying the proof is that each $S_i(K_S)$ is just a copy of the set K_S scaled by a factor c_i, and so there should be a relationship between the number of boxes of side $r > 0$ that intersect K_S, and the number of boxes of side $c_i r$ that intersect $S_i(K_S)$.

If we look at a grid G_r and its image under a similarity S_i, then we obtain a grid of boxes of side-length $c_i r$. Hence, if $N_r(K_S)$ boxes from G_r meet K_S, then the same number of boxes from the transformed grid meet $S_i(K_S)$. We would like to conclude that $N_{c_i r}(S_i(K_S)) = N_r(K_S)$, but unfortunately, the image of a box from G_r may not be a box from the grid $G_{c_i r}$ (as indicated in Figure 3.10), and so the equality need not hold. However, it is possible to show that there are real numbers $0 < a \leq 1 \leq b$ so that, for any non-empty compact set K and similarity S with similarity ratio c,

$$aN_r(K) \leq N_{cr}(S(K)) \leq bN_r(K). \tag{3.2}$$

Now

$$K_S = S_1(K_S) \cup S_2(K_S),$$

and if we assume that this union is disjoint, then there is a $\delta > 0$ such that for all $\mathbf{x} \in S_1(K_S)$ and $\mathbf{y} \in S_2(K_S)$,

$$d^{(2)}(\mathbf{x}, \mathbf{y}) \geq \delta.$$

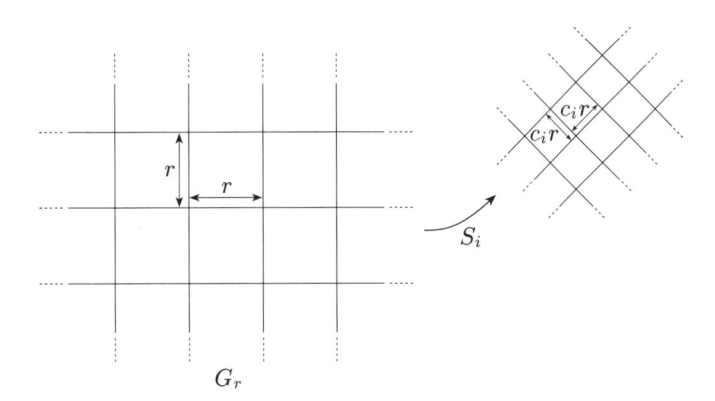

Figure 3.10

Hence if $d > 0$ is small enough, then a box in $G_d(K_S)$ can only intersect at most one of the sets $S_1(K_S)$ and $S_2(K_S)$, and so

$$N_d(K_S) = N_d(S_1(K_S)) + N_d(S_2(K_S)). \tag{3.3}$$

Thus, by using (3.2) with $K = K_S$, $S = S_i$ and $r = d/c_i$ for $i = 1, 2$, we find

$$a(N_{d/c_1}(K_S) + N_{d/c_2}(K_S)) \leq N_d(K_S) \leq b(N_{d/c_1}(K_S) + N_{d/c_2}(K_S)).$$

> In the full proof, the open set condition plays the role of disjointness and ensures that an equation similar to (3.3) is still true.

If we now suppose that $s = \dim K_S$ exists, and in particular that there is a positive number m such that

$$N_d(K_s) = m\left(\frac{1}{d}\right)^s \quad \text{for } 0 < d \leq 1, \tag{3.4}$$

then we deduce that, for sufficiently small $d > 0$,

$$amd^{-s}\left(c_1^s + c_2^s\right) \leq md^{-s} \leq bmd^{-s}\left(c_1^s + c_2^s\right).$$

Hence

$$a\left(c_1^s + c_2^s\right) \leq 1 \leq b\left(c_1^s + c_2^s\right).$$

Now consider the similarities $S_i \circ S_j$ for $i, j = 1, 2$. The similarity $S_i \circ S_j$ has similarity ratio $c_i c_j$, and

$$K_S = \bigcup_{i,j=1}^{2} (S_i \circ S_j)(K_S).$$

> The composition of two similarities is a similarity whose similarity ratio is the product of the original ratios.

Hence, by repeating the preceding argument for the four similarities $S_1 \circ S_1$, $S_1 \circ S_2$, $S_2 \circ S_1$, $S_2 \circ S_2$, we find

$$a((c_1c_1)^s + (c_1c_2)^s + (c_2c_1)^s + (c_2c_2)^s) \leq 1 \leq b((c_1c_1)^s + (c_1c_2)^s + (c_2c_1)^s + (c_2c_2)^s).$$

But

$$(c_1c_1)^s + (c_1c_2)^s + (c_2c_1)^s + (c_2c_2)^s = (c_1^s + c_2^s)^2,$$

and so

$$a\left(c_1^s + c_2^s\right)^2 \leq 1 \leq b\left(c_1^s + c_2^s\right)^2.$$

Similarly, for each $n \in \mathbb{N}$, by considering the similarities $S_{i_1} \circ S_{i_2} \circ \cdots \circ S_{i_n}$ for $1 \leq i_1, i_2, \ldots, i_n \leq 2$, we find that

$$a\left(c_1^s + c_2^s\right)^n \leq 1 \leq b\left(c_1^s + c_2^s\right)^n.$$

Thus

$$1/b \leq \left(c_1^s + c_2^s\right)^n \leq 1/a \quad \text{for each } n \in \mathbb{N}.$$

This is possible only if

$$c_1^s + c_2^s = 1,$$

as claimed.

The complete argument to show that the box dimension exists is based on similar ideas but requires more work — we omit the details. ∎

Worked problem 3.2

Let K_0 and K_1 be as shown below.

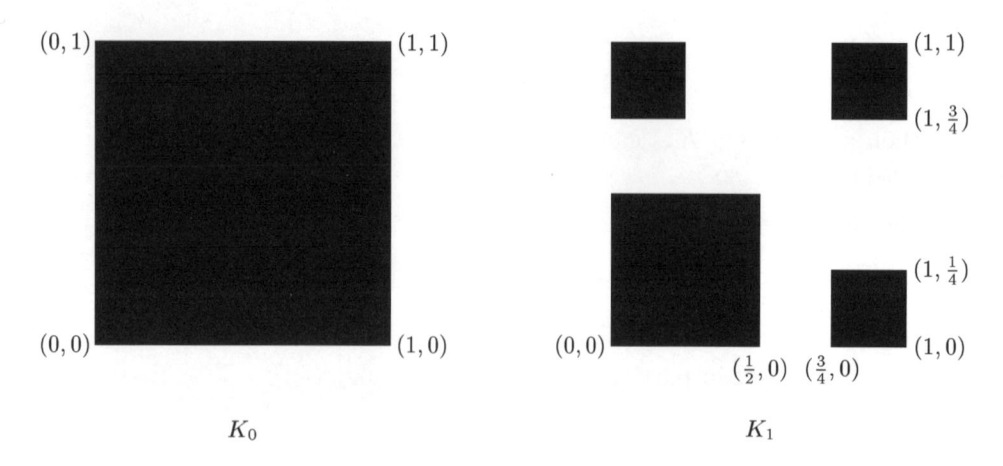

Figure 3.11

(a) Determine similarities S_1, S_2, S_3, S_4 such that $K_1 = \bigcup_{i=1}^4 S_i(K_0)$.
(b) Let $K_{n+1} = \bigcup_{i=1}^4 S_i(K_n)$, for each $n \in \mathbb{N}$. Sketch the set K_2.
(c) Show that $K = \lim_{n \to \infty} K_n$ exists and is the invariant set of S_1, S_2, S_3, S_4.
(d) Show that S_1, S_2, S_3, S_4 satisfy the open set condition.
(e) Determine $\dim K$.

Solution

(a) A natural choice of similarities is

$$S_1(x,y) = \left(\tfrac{1}{2}x, \tfrac{1}{2}y\right), \qquad S_2(x,y) = \left(\tfrac{1}{4}x + \tfrac{3}{4}, \tfrac{1}{4}y\right),$$
$$S_3(x,y) = \left(\tfrac{1}{4}x + \tfrac{3}{4}, \tfrac{1}{4}y + \tfrac{3}{4}\right), \qquad S_4(x,y) = \left(\tfrac{1}{4}x, \tfrac{1}{4}y + \tfrac{3}{4}\right).$$

(b)

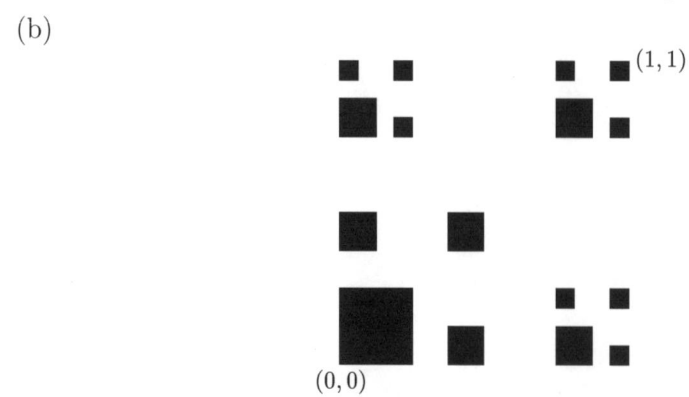

Figure 3.12

(c) Let $S\colon \mathcal{K}(\mathbb{R}^2) \to \mathcal{K}(\mathbb{R}^2)$ be defined by

$$S(K) = \bigcup_{i=1}^{4} S_i(K).$$

Then $K_{n+1} = S(K_n)$, and hence $K_n = S^n(K_0)$ for each $n \in \mathbb{N}$. Since K_0 is non-empty and compact, it follows from Theorem 2.8 that $K = \lim_{n \to \infty} S^n(K_0)$ exists and is the invariant set of S_1, S_2, S_3, S_4.

(d) The following diagram shows that S_1, S_2, S_3, S_4 satisfy the open set condition.

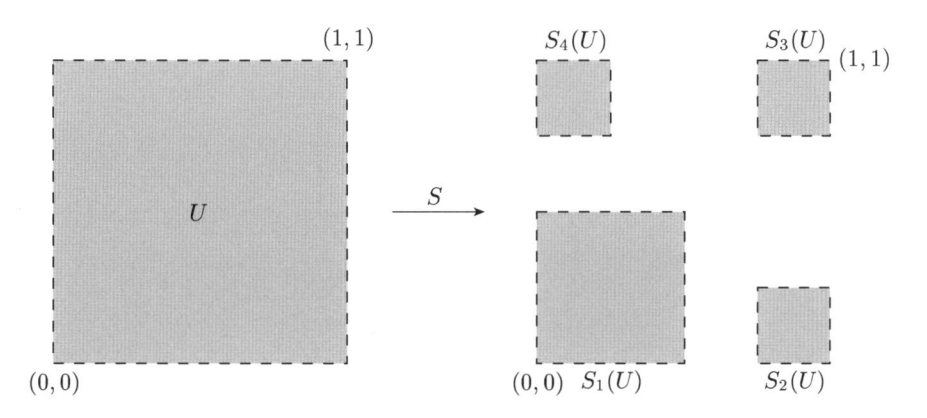

Figure 3.13

(e) It follows from parts (c) and (d) that the conditions of Theorem 3.1 are satisfied. Since S_1 has similarity ratio $\frac{1}{2}$ and each of the similarities S_2, S_3 and S_4 has similarity ratio $\frac{1}{4}$, the box dimension of K is the number s for which

$$\left(\tfrac{1}{2}\right)^s + 3\left(\tfrac{1}{4}\right)^s = 1.$$

By substituting $x = \left(\tfrac{1}{2}\right)^s$, we obtain

$$x + 3x^2 = 1,$$

which has solutions $x = \tfrac{1}{6}(-1 + \sqrt{13})$ and $x = \tfrac{1}{6}(-1 - \sqrt{13})$.

But x must be positive, and so

$$\left(\tfrac{1}{2}\right)^s = \tfrac{1}{6}(-1 + \sqrt{13}).$$

Thus

$$\dim K = s = \frac{\log \tfrac{1}{6}(-1 + \sqrt{13})}{-\log 2} = \frac{\log 6 - \log(-1 + \sqrt{13})}{\log 2}$$

$$\simeq 1.203\ldots. \qquad\blacksquare$$

The formula for the dimension of a set satisfying the hypotheses of Theorem 3.1 is particularly simple when all the similarities involved have the same similarity ratio — this is the case for each of the fractals in Section 1. For suppose that K_S is the invariant set of the similarities $S_1, S_2, \ldots, S_k$, each with similarity ratio $c \in (0, 1)$, and that $S_1, S_2, \ldots, S_k$ satisfy the open set condition. It follows from Theorem 3.1 that $\dim K_S = s$, where s satisfies

$$\sum_{i=1}^{k} c^s = kc^s = 1,$$

so

$$k = c^{-s}.$$

Taking logarithms, we obtain

$$\log k = -s \log c,$$

that is,

$$s = \frac{\log k}{-\log c}.$$

We have proved the following result.

Theorem 3.2

Let K_S be the invariant set of the similarities $S_1, S_2, \ldots, S_k$, each with similarity ratio $c \in (0,1)$. If $S_1, S_2, \ldots, S_k$ satisfy the open set condition, then

$$\dim K_S = \frac{\log k}{-\log c}.$$

This theorem gives a simple way of calculating the box dimension of a large class of fractals.

Sierpiński gasket

The Sierpiński gasket K_G is the invariant set of the similarities

$$S_1(x,y) = \left(\tfrac{1}{2}x, \tfrac{1}{2}y\right), \qquad S_2(x,y) = \left(\tfrac{1}{2}x + \tfrac{1}{2}, \tfrac{1}{2}y\right),$$

$$S_3(x,y) = \left(\tfrac{1}{2}x + \tfrac{1}{4}, \tfrac{1}{2}y + \tfrac{\sqrt{3}}{4}\right).$$

See Problem 1.7.

In Problem 3.3 you showed that S_1, S_2 and S_3 satisfy the open set condition. Since each of these similarities has similarity ratio $\tfrac{1}{2}$, it follows from Theorem 3.2 that

$$\dim K_G = \frac{\log 3}{-\log \tfrac{1}{2}} = \frac{\log 3}{\log 2} \simeq 1.584 \ldots .$$

So the dimension of the Sierpiński gasket lies between 1 and 2, as we predicted.

von Koch curve

The von Koch curve K_V is the invariant set of four similarities.

S_1 scales by a factor of $\tfrac{1}{3}$.

S_2 scales by a factor of $\tfrac{1}{3}$, rotates anticlockwise by $\tfrac{1}{3}\pi$ about $(0,0)$ and translates by $\left(\tfrac{1}{3}, 0\right)$.

S_3 scales by a factor of $\tfrac{1}{3}$, rotates clockwise by $\tfrac{1}{3}\pi$ about the point $(0,0)$, and translates by $\left(\tfrac{1}{2}, \tfrac{\sqrt{3}}{6}\right)$.

S_4 scales by a factor of $\tfrac{1}{3}$ and translates by $\left(\tfrac{2}{3}, 0\right)$.

Figure 3.14 shows that these similarities satisfy the open set condition.

Here, U is the interior of the triangle based on $[0,1]$ with height $\tfrac{\sqrt{3}}{6}$.

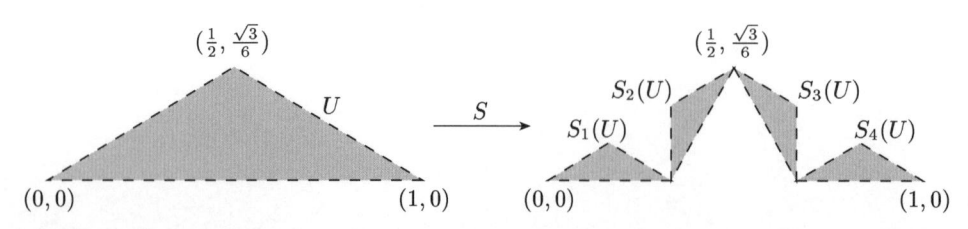

Figure 3.14

Since each of these four similarities has similarity ratio $\frac{1}{3}$, it follows from Theorem 3.2 that
$$\dim K_V = \frac{\log 4}{-\log \frac{1}{3}} = \frac{\log 4}{\log 3} \simeq 1.261\ldots.$$

Thus the dimension of the von Koch curve lies between 1 and 2, as we expected.

So, the three examples of fractals that you met in Section 1 — the middle-third Cantor set, the Sierpiński gasket and the von Koch curve — all satisfy the conditions of Theorem 3.2 and their dimensions can be calculated easily.

We now look at some other examples of fractals for which this is true.

Worked problem 3.3

Let K_0 and K_1 be the curves shown below.

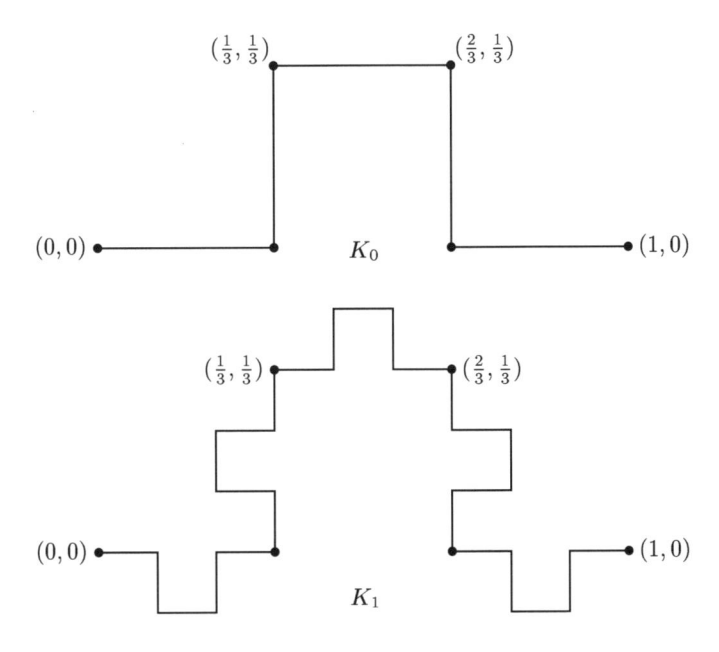

Figure 3.15

(a) Determine similarities $S_1, S_2, \ldots, S_5$ such that $K_1 = \bigcup_{i=1}^{5} S_i(K_0)$.
(b) Let $K_{n+1} = \bigcup_{i=1}^{5} S_i(K_n)$, for each $n \in \mathbb{N}$. Sketch the set K_2.
(c) Show that $K = \lim_{n \to \infty} K_n$ exists and is the invariant set of $S_1, S_2, \ldots, S_5$.
(d) Show that $S_1, S_2, \ldots, S_5$ satisfy the open set condition.
(e) Determine $\dim K$.

Solution

(a) A natural choice of similarities is
$$S_1(x,y) = \left(\tfrac{1}{3}x, -\tfrac{1}{3}y\right), \qquad S_2(x,y) = \left(-\tfrac{1}{3}y + \tfrac{1}{3}, \tfrac{1}{3}x\right),$$
$$S_3(x,y) = \left(\tfrac{1}{3}x + \tfrac{1}{3}, \tfrac{1}{3}y + \tfrac{1}{3}\right), \qquad S_4(x,y) = \left(\tfrac{1}{3}y + \tfrac{2}{3}, -\tfrac{1}{3}x + \tfrac{1}{3}\right),$$
$$S_5(x,y) = \left(\tfrac{1}{3}x + \tfrac{2}{3}, -\tfrac{1}{3}y\right).$$

(b)

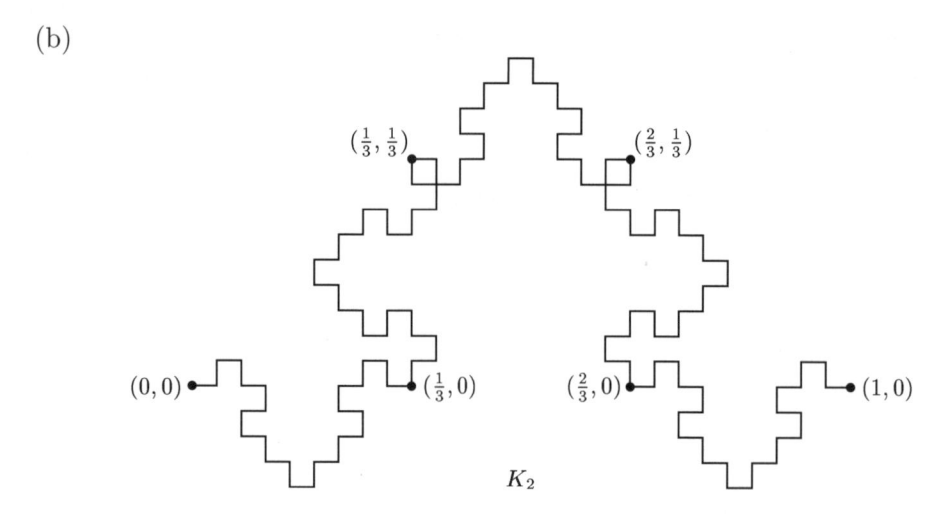

Figure 3.16

(c) Let $S: \mathcal{K}(\mathbb{R}^2) \to \mathcal{K}(\mathbb{R}^2)$ be defined by

$$S(K) = \bigcup_{i=1}^{5} S_i(K).$$

Then $K_{n+1} = S(K_n)$, and hence $K_n = S^n(K_0)$ for each $n \in \mathbb{N}$. Since K_0 is non-empty and compact, it follows from Theorem 2.8 that $K = \lim_{n \to \infty} S^n(K_0)$ exists and is the invariant set of $S_1, S_2, \ldots, S_5$.

(d) The following diagram shows that $S_1, S_2, \ldots, S_5$ satisfy the open set condition.

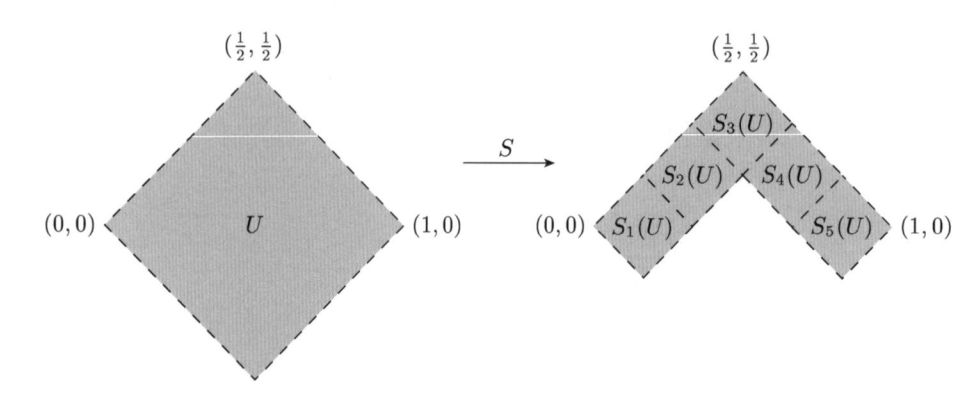

Figure 3.17

(e) It follows from parts (c) and (d) that the conditions of Theorem 3.2 are satisfied. Since each of the similarities $S_1, S_2, \ldots, S_5$ has similarity ratio $\frac{1}{3}$, we have

$$\dim K = \frac{\log 5}{-\log \frac{1}{3}} = \frac{\log 5}{\log 3} \simeq 1.464 \ldots. \qquad \blacksquare$$

Problem 3.4

Let K_0 and K_1 be the sets shown below.

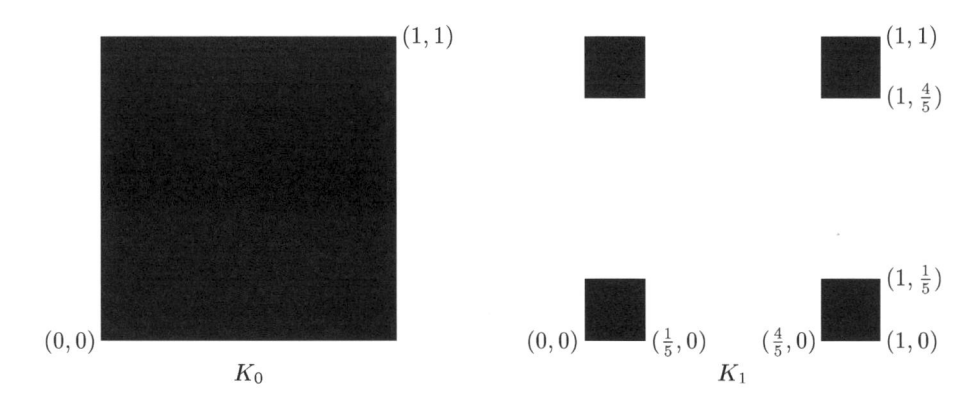

Figure 3.18

(a) Determine similarities S_1, S_2, S_3, S_4 such that $K_1 = \bigcup_{i=1}^{4} S_i(K_0)$.
(b) Let $K_{n+1} = \bigcup_{i=1}^{4} S_i(K_n)$, for each $n \in \mathbb{N}$. Sketch the set K_2.
(c) Show that $K = \lim_{n \to \infty} K_n$ exists and is the invariant set of S_1, S_2, S_3, S_4.
(d) Show that S_1, S_2, S_3, S_4 satisfy the open set condition.
(e) Determine $\dim K$.

Conclusion

In this unit, we have seen how general results concerning topological spaces can be used to great effect in the study of fractals. In particular, given a family of similarities $S_1, S_2, \ldots, S_k$, we have seen how the Contraction Mapping Theorem enables us to construct a non-empty compact set that is invariant for these similarities. Many sets constructed in this way are fractals, and a good estimate of their size can be obtained by using the formula given in Theorem 3.1.

This is also the last unit of the course, and we hope that it has given you a flavour of how topological ideas and results can be applied to study other areas of mathematics.

We have travelled a long way from our initial study of continuous functions on the real line, and have gained a deeper understanding of the role that topology plays in modern mathematics. Topology is a vital part of every pure mathematician's toolkit.

Solutions to problems

1.1 The set K_n consists of 2^n line segments, each of length $\frac{1}{3^n}$. Their total length is $L_n = \frac{2^n}{3^n} = \left(\frac{2}{3}\right)^n$, and $\lim_{n\to\infty} L_n = 0$.

1.2 (a) S has similarity ratio $\frac{1}{2}$.

(b) S has similarity ratio $\frac{1}{4}$.

1.3 There are many possibilities; for example, the similarities
$$S_1(x) = \tfrac{1}{2}x \quad \text{and} \quad S_2(x) = \tfrac{1}{2}(x+1)$$
have similarity ratio $\frac{1}{2}$ and
$$S_1([0,1]) \cup S_2([0,1]) = [0,\tfrac{1}{2}] \cup [\tfrac{1}{2},1] = [0,1].$$

1.4 (a) The set K_2 is shown below.

K_2

(b) The set K_n consists of 2^n line segments, each of length $\left(\frac{1}{4}\right)^n$. So the length of K_n is $L_n = \frac{2^n}{4^n} = \left(\frac{1}{2}\right)^n$, and $\lim_{n\to\infty} L_n = 0$.

(c) A natural choice of similarities is
$$S_1(x) = \tfrac{1}{4}x \quad \text{and} \quad S_2(x) = \tfrac{1}{4}x + \tfrac{1}{2},$$
and each has similarity ratio $\frac{1}{4}$.

1.5 Each of the 3^n triangles that form K_n has sides of length $\left(\frac{1}{2}\right)^n$, so each of these triangles has boundary length $3\left(\frac{1}{2}\right)^n$. It follows that the total length of their boundaries is $3^n \times \frac{3}{2^n} = 3\left(\frac{3}{2}\right)^n$.

The boundary of each triangle belongs to the Sierpiński gasket, and so the length of the Sierpiński gasket is at least $3\left(\frac{3}{2}\right)^n$ for each $n \in \mathbb{N}$. Since $\left(3\left(\frac{3}{2}\right)^n\right)$ is a divergent sequence, the Sierpiński gasket does not have finite length.

1.6 The set K_n consists of 3^n triangles, each of area
$$\frac{1}{2} \times \frac{\sqrt{3}/2}{2^n} \times \frac{1}{2^n} = \frac{\sqrt{3}}{4^{n+1}}.$$
So the area of K_n is $\frac{3^n \sqrt{3}}{4^{n+1}} = \frac{\sqrt{3}}{4}\left(\frac{3}{4}\right)^n$. Since $K_G \subset K_n$, the area of K_G is at most $\frac{\sqrt{3}}{4}\left(\frac{3}{4}\right)^n$ for each $n \in \mathbb{N}$. Since $\lim_{n\to\infty} \frac{\sqrt{3}}{4}\left(\frac{3}{4}\right)^n = 0$, the Sierpiński gasket has zero area.

1.7 A natural choice of similarities is
$$S_1(x,y) = \left(\tfrac{1}{2}x, \tfrac{1}{2}y\right),$$
$$S_2(x,y) = \left(\tfrac{1}{2}x + \tfrac{1}{2}, \tfrac{1}{2}y\right),$$
$$S_3(x,y) = \left(\tfrac{1}{2}x + \tfrac{1}{4}, \tfrac{1}{2}y + \tfrac{\sqrt{3}}{4}\right),$$
and they each have similarity ratio $\frac{1}{2}$.

1.8 (a) The set K_2 is shown below.

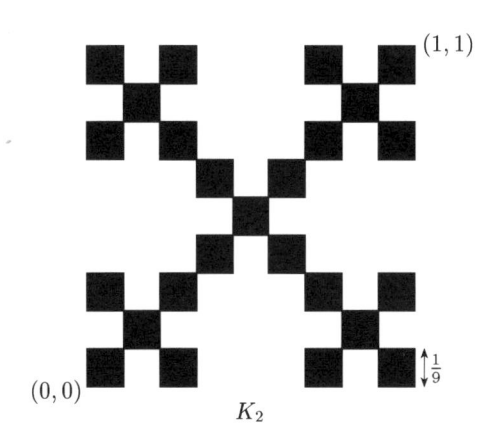

K_2

(b) Each set K_n is a finite union of closed bounded squares, and so is non-empty and compact. Since $K_{n+1} \subseteq K_n$, the set K is the intersection of a nested sequence of non-empty compact sets, and so is non-empty and compact.

The set K_n consists of 5^n closed squares, each with sides of length $\left(\frac{1}{3}\right)^n$ and area $\left(\frac{1}{9}\right)^n$, so the total area of K_n is $\left(\frac{5}{9}\right)^n$. Since $K \subset K_n$, the area of K is at most $\left(\frac{5}{9}\right)^n$ for each $n \in \mathbb{N}$. Since $\lim_{n\to\infty}\left(\frac{5}{9}\right)^n = 0$, K has zero area.

(c) A natural choice of similarities is
$$S_1(x,y) = \left(\tfrac{1}{3}x, \tfrac{1}{3}y\right),$$
$$S_2(x,y) = \left(\tfrac{1}{3}x + \tfrac{2}{3}, \tfrac{1}{3}y\right),$$
$$S_3(x,y) = \left(\tfrac{1}{3}x, \tfrac{1}{3}y + \tfrac{2}{3}\right),$$
$$S_4(x,y) = \left(\tfrac{1}{3}x + \tfrac{2}{3}, \tfrac{1}{3}y + \tfrac{2}{3}\right),$$
$$S_5(x,y) = \left(\tfrac{1}{3}x + \tfrac{1}{3}, \tfrac{1}{3}y + \tfrac{1}{3}\right),$$
and each has similarity ratio $\frac{1}{3}$.

1.9 The set K_1 consists of four line segments of length $\frac{1}{3}$. When constructing K_n, each line segment in K_{n-1} is replaced by four line segments each of $\frac{1}{3}$ of the length. So K_n is a connected curve consisting of 4^n line segments each of length $\left(\frac{1}{3}\right)^n$. Using the method given in the question, we conclude that K_V has length at least $\left(\frac{4}{3}\right)^n$, for each $n \in \mathbb{N}$. Since $\left(\left(\frac{4}{3}\right)^n\right)$ is a divergent sequence, K_V does not have finite length.

2.1 (a) $(-1, 1]$ is not closed and so is not compact — so it does not belong to $\mathcal{K}(\mathbb{R})$.

(b) $[0, 2] \in \mathcal{K}(\mathbb{R})$.

(c) $[0, \infty)$ is not bounded and so is not compact — so it does not belong to $\mathcal{K}(\mathbb{R})$.

(d) $\{1\} \in \mathcal{K}(\mathbb{R})$.

(e) $[0,2] \cap [4,6] = \varnothing \notin \mathcal{K}(\mathbb{R})$.

2.2 (a)

$d^{(1)}(1, [2, 3]) = d^{(1)}(1, 2) = 1.$

(b)

$d^{(1)}(1, [0, 2]) = d^{(1)}(1, 1) = 0.$

(c)

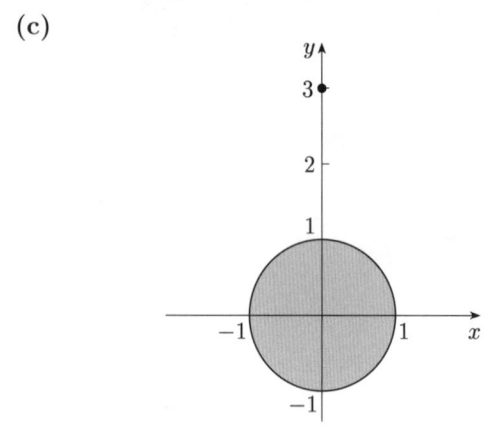

$d^{(2)}((0, 3), \{(x, y) : x^2 + y^2 \leq 1\})$
$= d^{(2)}((0, 3), (0, 1)) = 2.$

(d)

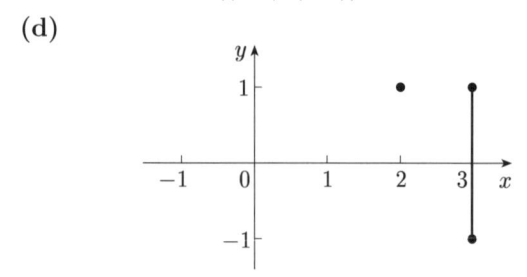

$d^{(2)}((2, 1), \{(3, y) : |y| \leq 1\}) = d^{(2)}((2, 1), (3, 1)) = 1.$

2.3 Let $J \subseteq K$. Then
$d^{(m)}(\mathbf{x}, K) = \min\{d^{(m)}(\mathbf{x}, \mathbf{k}) : \mathbf{k} \in K\}$
$\leq \min\{d^{(m)}(\mathbf{x}, \mathbf{k}) : \mathbf{k} \in J \cap K\}$
$= \min\{d^{(m)}(\mathbf{x}, \mathbf{k}) : \mathbf{k} \in J\} = d^{(m)}(\mathbf{x}, J).$

2.4 We begin by sketching the sets J and K.

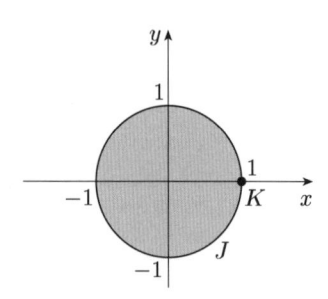

The point in J that is furthest from K is $(-1, 0)$, so
$d^{(2)}(J, K) = d^{(2)}((-1, 0), K)$
$= d^{(2)}((-1, 0), (1, 0)) = 2.$

There is only one point in K, so
$d^{(2)}(K, J) = d^{(2)}((1, 0), J) = d^{(2)}((1, 0), (1, 0)) = 0.$

2.5 (a) $d^{(m)}(J, K) = \max\{d^{(m)}(\mathbf{x}, K) : \mathbf{x} \in J\}$
$\leq \max\{d^{(m)}(\mathbf{x}, J) : \mathbf{x} \in J\},$

by Lemma 2.2, since $J \subseteq K$.
By Lemma 2.1, $d^{(m)}(\mathbf{x}, J) = 0$ if $\mathbf{x} \in J$, and so
$d^{(m)}(J, K) = 0.$

(b) $d^{(m)}(L, K) = \max\{d^{(m)}(\mathbf{x}, K) : \mathbf{x} \in L\}$
$\leq \max\{d^{(m)}(\mathbf{x}, J) : \mathbf{x} \in L\}$, by Lemma 2.2
$= d^{(m)}(L, J).$

2.6 The solution to Problem 2.4 shows that $d^{(2)}(J, K) = 2$ and $d^{(2)}(K, J) = 0$. So
$d_H(J, K) = \max\{2, 0\} = 2.$

2.7 (a) $S([0, 2]) = S_1([0, 2]) \cup S_2([0, 2])$
$= [0, \tfrac{2}{3}] \cup [\tfrac{2}{3}, \tfrac{4}{3}] = [0, \tfrac{4}{3}].$

This set belongs to $\mathcal{K}(\mathbb{R})$.

(b) $S([-2, 2]) = S_1([-2, 2]) \cup S_2([-2, 2])$
$= [-\tfrac{2}{3}, \tfrac{2}{3}] \cup [0, \tfrac{4}{3}] = [-\tfrac{2}{3}, \tfrac{4}{3}].$

This set belongs to $\mathcal{K}(\mathbb{R})$.

2.8 (a)

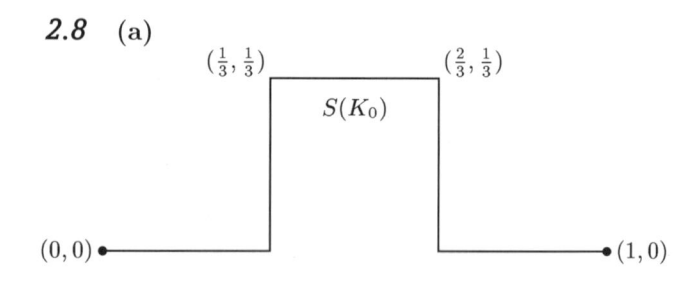

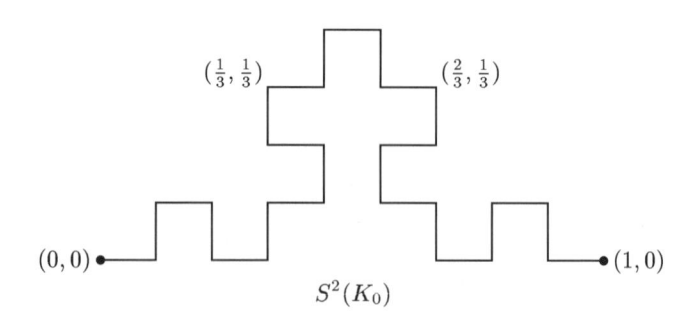

(b) Let $K_n = S^n(K_0)$, for each $n \in \mathbb{N}$.

The set K_1 consists of five line segments each of length $\tfrac{1}{3}$ and K_2 consists of 25 line segments each of length $(\tfrac{1}{3})^2$. In constructing K_n, each line segment of K_{n-1} is replaced by five line segments of one-third the length, so K_n consists of 5^n line segments of length $\tfrac{1}{3^n}$ and so has length $\left(\tfrac{5}{3}\right)^n$.

(c) Since $K_0 \in \mathcal{K}(\mathbb{R}^2)$, and S is a contraction mapping with contraction ratio $\tfrac{1}{3}$, it follows from Theorem 2.8 that there is a unique set $K_S \in \mathcal{K}(\mathbb{R}^2)$ such that $K_S = \lim_{n \to \infty} S^n(K_0)$.
(This set satisfies
$$K_S = S(K_S) = \bigcup_{i=1}^{5} S_i(K_S)$$
and so is self-similar.)

2.9 $S_1(\emptyset) = S_2(\emptyset) = \emptyset$, and so $S_1(\emptyset) \cup S_2(\emptyset) = \emptyset$.
$S_1(\mathbb{R}) = S_2(\mathbb{R}) = \mathbb{R}$, and so $S_1(\mathbb{R}) \cup S_2(\mathbb{R}) = \mathbb{R}$.
$S_1([0,\infty)) = [0,\infty)$ and $S_2([0,\infty)) = [\frac{2}{3},\infty)$ and so $S_1([0,\infty)) \cup S_2([0,\infty)) = [0,\infty)$.

2.10 The set $[0,2]$ is closed and bounded, and hence is in $\mathcal{K}(\mathbb{R})$. Also,

$$S([0,2]) = [0, \tfrac{2}{3}] \cup [\tfrac{2}{3}, \tfrac{4}{3}] = [0, \tfrac{4}{3}] \subseteq [0,2].$$

It follows from Theorem 2.8 and Lemma 2.9 that $\bigcap_{n \in \mathbb{N}} S^n([0,2])$ is the unique set in $\mathcal{K}(\mathbb{R})$ that is invariant for S_1 and S_2. Thus $\bigcap_{n \in \mathbb{N}} S^n([0,2]) = K_C$.

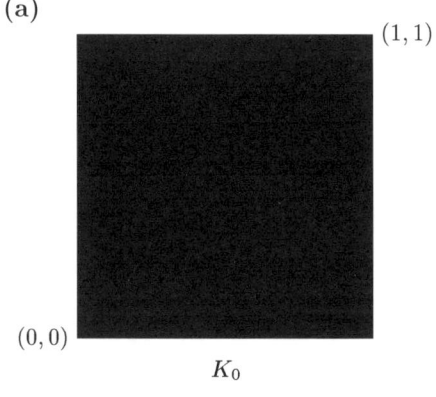

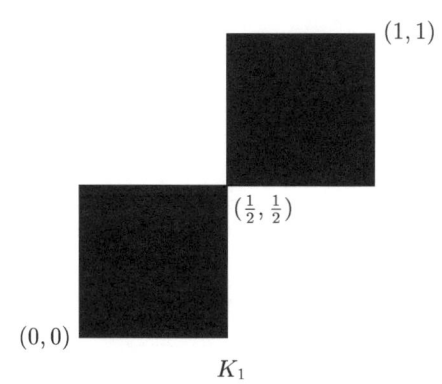

2.11 (a)

K_0: square with corners $(0,0)$ and $(1,1)$.

K_1: two squares, one with corners $(0,0)$ and $(\frac{1}{2}, \frac{1}{2})$, the other with corners $(\frac{1}{2}, \frac{1}{2})$ and $(1,1)$.

(b) It follows from Theorem 2.8 that $\lim_{n \to \infty} K_n$ is the unique set in $\mathcal{K}(\mathbb{R}^2)$ that is invariant for S. Now if $K' = \{(x,y) : y = x, 0 \le x \le 1\}$, then

$$\begin{aligned} S(K') &= S_1(K') \cup S_2(K') \\ &= \{(x,y) : y = x, 0 \le x \le \tfrac{1}{2}\} \\ &\quad \cup \{(x,y) : y = x, \tfrac{1}{2} \le x \le 1\} \\ &= K'. \end{aligned}$$

Thus K' is invariant for S.

Since $K' \in \mathcal{K}(\mathbb{R}^2)$, we deduce that $K' = \lim_{n \to \infty} K_n$.

3.1 (a)

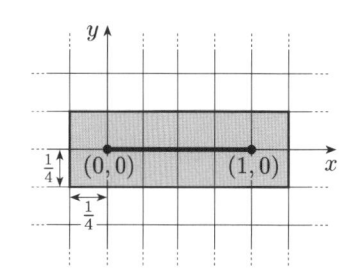

$N_{\frac{1}{4}}(A) = 12.$

(b)

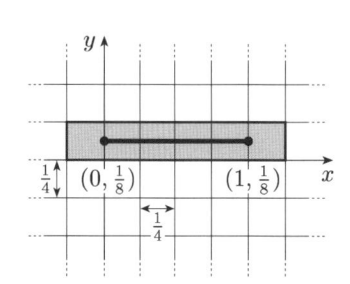

$N_{\frac{1}{4}}(A) = 6.$

(c)

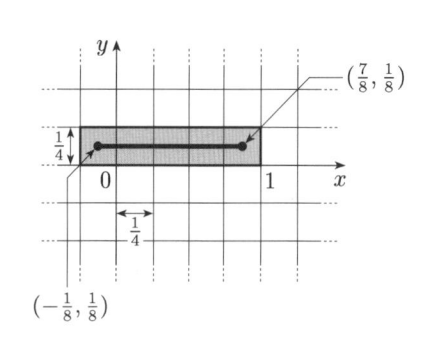

$N_{\frac{1}{4}}(A) = 5.$

(d)

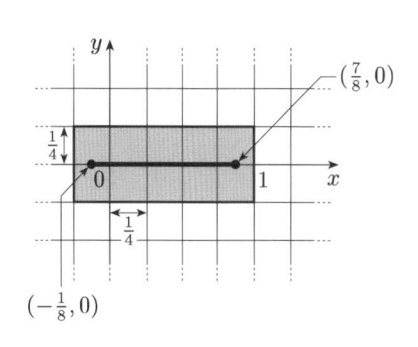

$N_{\frac{1}{4}}(A) = 10.$

3.2 We follow the technique described in Worked problem 3.1.

Our objective is to find
$$\lim_{d\to 0} \frac{\log N_d(K_S)}{-\log d}.$$

For a given $d > 0$, we estimate $N_d(K_S)$ by comparing it with $N_d(K_n)$, where n is chosen so that 4^{-n} is approximately equal to d.

We note first that, at the nth stage of the construction of K_S, the set K_n consists of 2^n line segments each of length $\left(\frac{1}{4}\right)^n$. The set K_n contains K_S, and since the endpoints of the intervals making up K_n are not removed in the construction, each of the 2^n intervals of K_n contains a point of K_S.

Let $d < 1$ and choose an $n \in \mathbb{N}$ so that
$$4^{-n} \leq d < 4^{-(n-1)}.$$

We find an upper bound for $N_d(K_S)$ by observing that each of the 2^n line segments making up K_n intersects at most six boxes from G_d, and so, since $K_S \subseteq K_n$,
$$N_d(K_S) \leq N_d(K_n) \leq 6 \times 2^n = 3 \times 2^{n+1}.$$

To find a lower bound for $N_d(K_S)$, we notice that each interval in K_n contains a point of K_S, and each box in G_d can intersect at most two intervals in K_n, so
$$N_d(K_S) \geq \frac{2^n}{2} = 2^{n-1}.$$

Hence
$$2^{n-1} \leq N_d(K_S) \leq 3 \times 2^{n+1},$$

and so
$$(n-1)\log 2 \leq \log N_d(K_C)$$
$$\leq (n+1)\log 2 + \log 3. \qquad (S.1)$$

Now, since $4^{-n} \leq d < 4^{-(n-1)}$,
$$-n\log 4 \leq \log d < -(n-1)\log 4,$$

and so, for $n \geq 2$,
$$\frac{1}{n\log 4} \leq \frac{1}{-\log d} < \frac{1}{(n-1)\log 4}.$$

Thus, for $n \geq 2$ and $4^{-n} \leq d \leq 4^{-(n-1)}$, we can combine these inequalities with (S.1) to obtain
$$\frac{(n-1)\log 2}{n\log 4} \leq \frac{\log N_d(K_S)}{-\log d} < \frac{(n+1)\log 2 + \log 3}{(n-1)\log 4}.$$

Now
$$\lim_{n\to\infty} \frac{(n-1)\log 2}{n\log 4} = \lim_{n\to\infty} \left(\frac{1-(1/n)}{1}\right) \frac{\log 2}{\log 4}$$
$$= \frac{\log 2}{\log 4} = \frac{1}{2}$$

and
$$\lim_{n\to\infty} \frac{(n+1)\log 2 + \log 3}{(n-1)\log 4}$$
$$= \lim_{n\to\infty} \left(\frac{1+(1/n)}{1-(1/n)}\right) \frac{\log 2}{\log 4} + \lim_{n\to\infty}\left(\frac{1}{n-1}\right)\frac{\log 3}{\log 4}$$
$$= \frac{\log 2}{\log 4} = \frac{1}{2}.$$

Since $n \to \infty$ as $d \to 0$, we deduce
$$\dim K_S = \lim_{d\to 0} \frac{\log N_d(K_C)}{-\log d} = \frac{1}{2}.$$

3.3 We take U to be the interior of the equilateral triangle based on the line segment joining $(0,1)$ to $(1,0)$, as shown below.

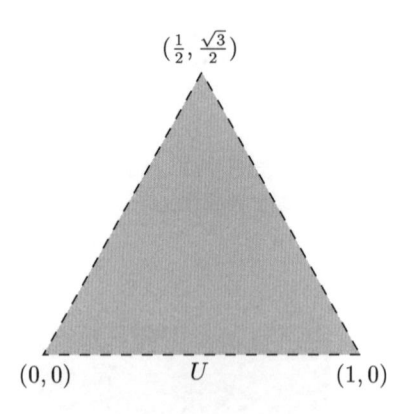

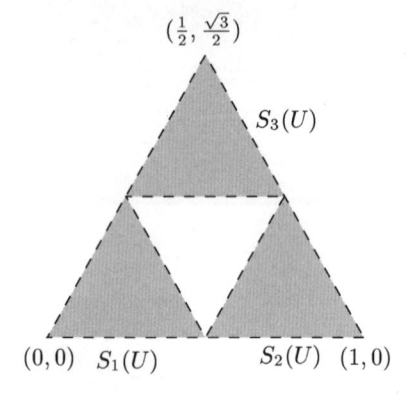

From the figure, we see that the sets $S_1(U)$, $S_2(U)$ and $S_3(U)$ are disjoint and $S_1(U) \cup S_2(U) \cup S_3(U) \subset U$, and so the open set condition is satisfied.

3.4 (a) A natural choice of similarities is
$$S_1(x,y) = \left(\tfrac{1}{5}x, \tfrac{1}{5}y\right),$$
$$S_2(x,y) = \left(\tfrac{1}{5}(x+4), \tfrac{1}{5}y\right),$$
$$S_3(x,y) = \left(\tfrac{1}{5}x, \tfrac{1}{5}(y+4)\right),$$
$$S_4(x,y) = \left(\tfrac{1}{5}(x+4), \tfrac{1}{5}(y+4)\right).$$

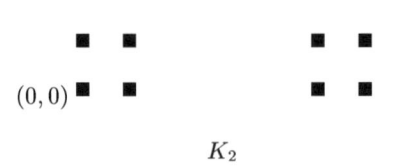

(b)

(c) Let $S\colon \mathcal{K}(\mathbb{R}^2) \to \mathcal{K}(\mathbb{R}^2)$ be defined by
$$S(K) = \bigcup_{i=1}^{4} S_i(K).$$
Then $K_{n+1} = S(K_n)$, and hence $S_n = S^n(K_0)$ for each $n \in \mathbb{N}$. Since K_0 is non-empty and compact, it follows from Theorem 2.8 that $K = \lim_{n \to \infty} S^n(K_0)$ exists and is the invariant set of S_1, S_2, S_3, S_4.

(d) The diagram below shows that S_1, S_2, S_3, S_4 satisfy the open set condition.

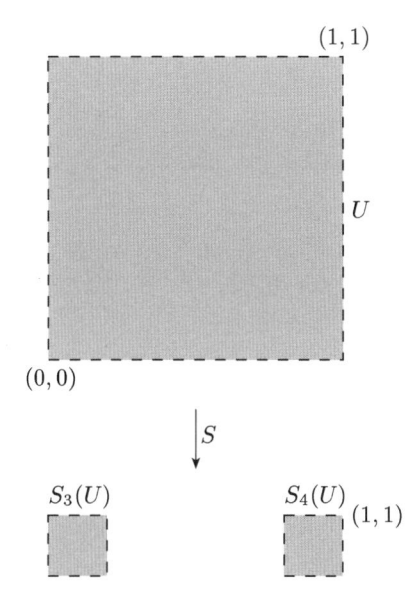

(e) It follows from parts (c) and (d) that the conditions of Theorem 3.2 are satisfied. Since each of the similarities S_1, S_2, S_3, S_4 has similarity ratio $\frac{1}{5}$, we have
$$\dim K = \frac{\log 4}{-\log \frac{1}{5}} = \frac{\log 4}{\log 5}.$$

Index

box dimension, 33, 35

Cantor sets, 5, 27
   construction of, 5, 27
   middle-third, 5, 27
contraction mapping, 23
Contraction Mapping Theorem, 16
contraction ratio, 23

d_H, Hausdorff distance, 20
dimension, 4, 33, 35
distance
   Euclidean, 17, 18
   Hausdorff, 20

Euclidean distance
   between two compact sets, 18
   from a point to a compact set, 17

fractal, 4

G_d, 33
grid, 33

Hausdorff distance, 20

invariant (for mappings), 25
iterated function scheme, 25

K_C, 6
K_G, 9
$\mathcal{K}(\mathbb{R}^m)$, 17
   completeness, 21
K_V, 12, 26

middle-third Cantor set, 5, 27

$N_d(A)$, 33

open set condition, 37
   and self-similar sets, 38, 42

self-similar, 8
   and the open set condition, 38, 42
Sierpiński gasket, 9, 29, 42
   construction of, 9, 29
similarity, 7
similarity ratio, 7

von Koch curve, 11, 26, 42
   construction of, 11, 26
von Koch snowflake, 4